Open Group Standard

CW00497400

ArchiMate® 2.0 Specification

The Open Group Publications available from Van Haren Publishing

The TOGAF Series:
TOGAF® Version 9.1
TOGAF® Version 9.1 - A Pocket Guide
TOGAF® 9 Foundation Study Guide, 2nd Edition
TOGAF® 9 Certified Study Guide, 2nd Edition

The Open Group Series:
Cloud Computing for Business – The Open Group Guide
ArchiMate 2.0 - A Pocket Guide
ArchiMate® 2.0 Specification

The Open Group Security Series:
Open Information Security Management Maturity Model (O-ISM3)
Open Enterprise Security Architecture (O-ESA)
Risk Management - The Open Group Guide

All titles are available to purchase from:
www.opengroup.org
www.vanharen.net
and also many international and online distributors.

Open Group Standard

ArchiMate® 2.0 Specification

Title: **ArchiMate® 2.0 Specification**

A Publication of: The Open Group

Publisher: Van Haren Publishing, Zaltbommel, www.vanharen.net

ISBN Hard copy: 978 90 8753 692 3
ISBN eBook: 978 90 8753 947 4

Edition: First edition, second impression, June 2012
First edition, second impression, March 2013

Design and Layout: CO2 Premedia, Amersfoort-NL

Document Number: C118

Comments relating to the material contained in this document may be submitted to:
The Open Group
Apex Plaza
Forbury Road
Reading
Berkshire, RG1 1AX
United Kingdom
or by electronic mail to: ogspecs@opengroup.org

Contents

Table of Figures

Preface

The Open Group

The Open Group is a global consortium that enables the achievement of business objectives through IT standards. With more than 400 member organizations, The Open Group has a diverse membership that spans all sectors of the IT community – customers, systems and solutions suppliers, tool vendors, integrators, and consultants, as well as academics and researchers – to:

- Capture, understand, and address current and emerging requirements, and establish policies and share best practices
- Facilitate interoperability, develop consensus, and evolve and integrate specifications and open source technologies
- Offer a comprehensive set of services to enhance the operational efficiency of consortia
- Operate the industry's premier certification service

Further information on The Open Group is available at www.opengroup.org.

The Open Group publishes a wide range of technical documentation, most of which is focused on development of Open Group Standards and Guides, but which also includes white papers, technical studies, certification and testing documentation, and business titles. Full details and a catalog are available at www.opengroup.org/bookstore.

Readers should note that updates – in the form of Corrigenda – may apply to any publication. This information is published at www.opengroup.org/corrigenda.

This Document

This document is The Open Group Standard for the ArchiMate 2.0 Specification.

Issue 2.0 includes a number of corrections, clarifications, and improvements compared to the previous issue, as well as two optional language extensions: the Motivation extension and the Implementation and Migration extension.

Intended Audience

The intended audience of this Technical Standard is threefold:

- Enterprise architecture practitioners, such as architects (application, information, process, infrastructure, products/services, and, obviously, enterprise architects), senior and operational management, project leaders, and anyone committed to work within the reference framework defined by the enterprise architecture. It is assumed that the reader has a certain skill level and is effectively committed to enterprise architecture. Such a person is most likely the architect – that is, someone who has affinity with modeling techniques, knows his way around the organization, and is familiar with information technology.
- Those who intend to implement ArchiMate in a software tool. They will find a complete and detailed description of the language in this document.
- The academic community, on which we rely for amending and improving the language based on state-of-the-art research results in the architecture field.

Structure

The structure of this Technical Standard is as follows:

- Chapter 1, Introduction, provides a brief introduction to the purpose of this standard.
- Chapter 2, Language Structure, presents some general ideas, principles, and assumptions underlying the development of the ArchiMate metamodel and introduces the ArchiMate framework.
- Chapter 3, Business Layer, covers the definition and usage of the business layer concept, together with examples.
- Chapter 4, Application Layer, covers the definition and usage of the application layer concept, together with examples.
- Chapter 5, Technology Layer, covers the definition and usage of the technical infrastructure layer concept, together with examples.
- Chapter 6, Cross-Layer Dependencies, and Chapter 7, Relationships, cover the definition of relationship concepts in a similar way.
- Chapter 8, Architecture Viewpoints, presents and clarifies a set of architecture viewpoints, developed in ArchiMate based on practical experience. All ArchiMate viewpoints are described in detail. For each viewpoint the comprised concepts and relationships, the guidelines for the viewpoint use, and the goal and target group and of the viewpoint

are specified. Furthermore, each viewpoint description contains example models.

- Chapter 9, Language Extension Mechanisms, handles extending and/or specializing the ArchiMate language for specialized or domain-specific purposes.
- Chapter 10, Motivation Extension, describes an optional language extension with concepts, relationships, and viewpoints for expressing the motivation for an architecture (e.g., stakeholders, concerns, goals, principles, and requirements).
- Chapter 11, Implementation and Migration Extension, describes an optional language extension with concepts, relationships, and viewpoints for expressing the implementation and migration aspects of an architecture (e.g., project, programs, plateaus, and gaps).
- Chapter 12, Future Directions, is an informative chapter that identifies extensions and directions for developments in the next versions of the language.

Trademarks

Boundaryless Information Flow™ is a trademark and ArchiMate®, Jericho Forum®, Making Standards Work®, Motif®, OSF/1®, The Open Group®, TOGAF®, UNIX®, and the "X" device are registered trademarks of The Open Group in the United States and other countries.

Java® is a registered trademark of Oracle and/or its affiliates.

MDA®, Model Driven Architecture®, OMG®, and UML® are registered trademarks and BPMN™, Business Process Modeling Notation™, MOF™, and Unified Modeling Language™ are trademarks of the Object Management Group..

All other brands, company, and product names are used for identification purposes only and may be trademarks that are the sole property of their respective owners.

Acknowledgements

The Open Group gratefully acknowledges the contribution of the following people in the development of this Open Group Standard:

- Maria-Eugenia Iacob, University of Twente
- Henk Jonkers, BiZZdesign BV
- Marc M. Lankhorst, Novay
- Erik (H.A.) Proper, Public Research Centre Henri Tudor & Radboud University Nijmegen
- Dick A.C. Quartel, BiZZdesign BV

The Open Group and ArchiMate project team would like to thank in particular the following individuals for their support and review of this Open Group Standard:

- Iver Band, Standard Insurance Company
- Mary Beijleveld, UWV
- Alexander Bielowski, Software AG
- Adrian Campbell, Ingenia Consulting
- John Coleshaw, QA Ltd.
- Jörgen Dahlberg, Biner Consulting
- Garry Doherty, The Open Group
- Wilco Engelsman, BiZZdesign BV
- Roland Ettema, Logica
- Henry M. Franken, BiZZdesign BV
- Kirk Hansen, Kirk Hansen Consulting
- Jos van Hillegersberg, University of Twente
- Andrew Josey, The Open Group
- Louw Labuschagne, Real IRM
- Veer Muchandi, Hewlett-Packard
- Bill Poole, JourneyOne
- Henk Volbeda, Sogeti
- Egon Willemsz, UWV

The results presented in this Open Group Standard have largely been produced during the ArchiMate project, and The Open Group gratefully acknowledges the contribution of the many people – former members of the project team – who have contributed to them.

The ArchiMate project comprised the following organizations:

- ABN AMRO
- Centrum voor Wiskunde en Informatica
- Dutch Tax and Customs Administration
- Leiden Institute of Advanced Computer Science
- Ordina
- Radboud Universiteit Nijmegen
- Stichting Pensioenfonds ABP
- Novay

Referenced Documents

The following documents are referenced in this Open Group Standard:

[1] ISO/IEC 42010:2007, Systems and Software Engineering – Recommended Practice for Architectural Description of Software-Intensive Systems, Edition 1.

[2] Enterprise Architecture at Work: Modeling, Communication, and Analysis, M.M. Lankhorst et al, Springer, 2005.

[3] Architecture Principles: The Cornerstones of Enterprise Architecture, D. Greefhorst, E. Proper, Springer, 2011.

[4] The Open Group Architecture Framework TOGAF, Version 9, 2009.

[5] A Framework for Information Systems Architecture, J.A. Zachman, IBM Systems Journal, Volume 26, No. 3, pp. 276–292, 1987.

[6] ITU Recommendation X.901 | ISO/IEC 10746-1:1998, Information Technology – Open Distributed Processing – Reference Model – Part 1: Overview, International Telecommunication Union, 1996.

[7] Unified Modeling Language: Infrastructure, Version 2.0 (formal/05-05-05), Object Management Group, March 2006.

[8] Extending and Formalizing the Framework for Information Systems Architecture, J.F. Sowa, J.A. Zachman,, IBM Systems Journal, Volume 31, No. 3, pp. 590-616, 1992.

[9] Enterprise Ontology: Theory and Methodology, J.L.G. Dietz, Springer, 2006.

[10] Unified Modeling Language: Superstructure, Version 2.0 (formal/05-07-04), Object Management Group, August 2005.

[11] A Business Process Design Language, H. Eertink, W. Janssen, P. Oude Luttighuis, W. Teeuw, C. Vissers, in Proceedings of the First World Congress on Formal Methods, Toulouse, France, September 1999.

[12] Enterprise Business Architecture: The Formal Link between Strategy and Results, R. Whittle, C.B. Myrick, CRC Press, 2004.

[13] Composition of Relations in Enterprise Architecture, R.v. Buuren, H. Jonkers, M.E. Iacob, P. Strating, in Proceedings of the Second International Conference on Graph Transformation, pp. 39–53, Edited by H. Ehrig et al, Rome, Italy, 2004.

[14] Viewpoints: A Framework for Integrating Multiple Perspectives in System Development, A. Finkelstein, J. Kramer, B. Nuseibeh, L. Finkelstein, M. Goedicke, in International Journal on Software

Engineering and Knowledge Engineering, Volume 2, No. 1, pp. 31–58, 1992.

[15] Viewpoints for Requirements Definition, G. Kotonya, I. Sommerville, IEE/BCS Software Engineering Journal, Volume 7, No. 6, pp. 375–387, November 1992.

[16] Paradigm Shift – The New Promise of Information Technology, D. Tapscott, A. Caston, New York: McGraw-Hill, 1993.

[17] The 4+1 View Model of Architecture, P.B. Kruchten, IEEE Software, Volume 12, No. 6, pp. 42–50, 1995.

[18] Model-Driven Architecture: Applying MDA to Enterprise Computing, D. Frankel, Wiley, 2003.

[19] Performance and Cost Analysis of Service-Oriented Enterprise Architectures, H. Jonkers, M. E. Iacob, in Global Implications of Modern Enterprise Information Systems: Technologies and Applications, Edited by A. Gunasekaran, IGI Global, 2009.

[20] Business Process Modeling Notation Specification (dtc/06-02-01), Object Management Group, February 2006.

[21] The Chaos Report, The Standish Group, 1994.

[22] No Silver Bullet: Essence and Accidents of Software Engineering, F.P. Brooks, IEEE Computer, 20(4):10–19, 1987.

[23] Managing Successful Programs, Office of Government Commerce (OGC), Stationery Office Books, 2007.

[24] Managing Successful Projects with PRINCE2 – 2009 Edition, Office of Government Commerce (OGC), Stationery Office Books, 2009.

[25] A Guide to the Project Management Body of Knowledge (PMBoK Guide), Fourth Edition, Project Management Institute, 2009.

Chapter 1

Introduction

An architecture is typically developed because key people have concerns that need to be addressed by the business and IT systems within the organization. Such people are commonly referred to as the "stakeholders" in the system. The role of the architect is to address these concerns, by identifying and refining the requirements that the stakeholders have, developing views of the architecture that show how the concerns and the requirements are going to be addressed, and by showing the trade-offs that are going to be made in reconciling the potentially conflicting concerns of different stakeholders. Without the architecture, it is unlikely that all the concerns and requirements will be considered and met.

Architecture descriptions are formal descriptions of an information system, organized in a way that supports reasoning about the structural and behavioral properties of the system and its evolution. They define the components or building blocks that make up the overall information system, and provide a plan from which products can be procured, and subsystems developed, that will work together to implement the overall system. It thus enables you to manage your overall IT investment in a way that meets the needs of your business.

To provide a uniform representation for diagrams that describe enterprise architectures, the ArchiMate enterprise architecture modeling language has been developed. It offers an integrated architectural approach that describes and visualizes the different architecture domains and their underlying relations and dependencies.

ArchiMate is a lightweight and scalable language in several respects:

- Its architecture framework is simple but comprehensive enough to provide a good structuring mechanism for architecture domains, layers, and aspects.
- The language incorporates the concepts of the "service orientation" paradigm that promotes a new organizing principle in terms of (business,

application, and infrastructure) services for organizations, with far-reaching consequences for their enterprise architecture.

The role of the ArchiMate standard is to provide a graphical language for the representation of enterprise architectures over time (i.e., including transformation and migration planning), as well as their motivation and rationale. The evolution of the standard is closely linked to the developments of the TOGAF standard and the emerging results from The Open Group forums and work groups active in this area. As a consequence, the ArchiMate standard does not provide its own set of defined terms, but rather follows those provided by the TOGAF standard.

This is Issue 2.0 of the Technical Standard, which contains a number of corrections, improvements, and clarifications in the description of the core language as described in Issue 1.0, as well as two optional extensions of the language: the Motivation extension and the Implementation and Migration extension.

This specification contains the formal definition of ArchiMate as a visual design language with adequate concepts for specifying inter-related architectures, and specific viewpoints for selected stakeholders. This is complemented by some considerations regarding language extension mechanisms, analysis, and methodological support. Furthermore, this document is accompanied by a separate document, in which certification and governance procedures surrounding the specification are specified.

Chapter 2

Language Structure

The unambiguous specification and description of enterprise architecture's components and especially of their relationships requires an architecture modeling language that addresses the issue of consistent alignment and facilitates a coherent modeling of enterprise architectures.

This chapter presents the construction of the ArchiMate architecture modeling language. The precise definition and illustration of its generic set of core concepts and relationships follow in Chapters 3, 4, 5, 6 and 7. The concepts and relationships of the two language extensions are described in more detail in Chapters 10 and 11. They provide a proper basis for visualization, analysis, tooling, and use of these concepts and relationships.

Sections 2.1 through 2.5 discuss some general ideas, principles, and assumptions underlying the development of the ArchiMate metamodel. Section 2.6 presents the ArchiMate framework, which is used in the remainder of this document as a reference taxonomy scheme for architecture concepts, models, viewpoints, and views. Sections 2.7 and 2.8 describe the basic structure of the two language extensions. Section 2.9 briefly describes the relationship between ArchiMate and TOGAF.

2.1 Design Approach

A key challenge in the development of a general metamodel for enterprise architecture is to strike a balance between the specificity of languages for individual architecture domains, and a very general set of architecture concepts, which reflects a view of systems as a mere set of inter-related entities. Figure 1 illustrates that concepts can be described at different levels of specialization.

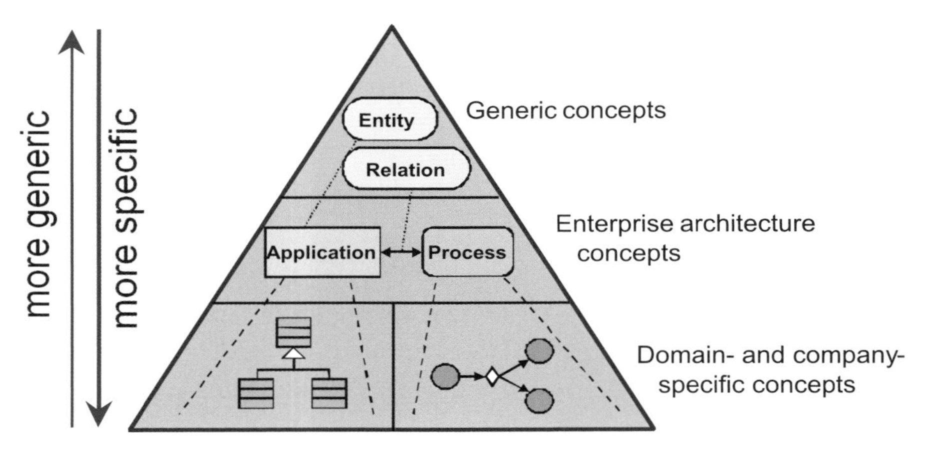

Figure 1: Metamodels at Different Levels of Specificity

At the base of the triangle we find the metamodels of the architecture modeling concepts used by specific organizations, as well as a variety of existing modeling languages and standards; UML is an example of a language in this category. At the top of the triangle we find the "most general" metamodel for system architectures, essentially a metamodel that merely comprises notions such as "entity" and "relation".

The design of the ArchiMate language started from a set of relatively generic concepts (higher up in the pyramid). These have been specialized towards application at different architectural layers, as explained below in the following sections.

The most important design restriction on the language is that it has been explicitly designed to be as small as possible, but still usable for most enterprise architecture modeling tasks. Many other languages, such as UML 2.0, try to accommodate all needs of all possible users. In the interest of simplicity of learning and use, ArchiMate has been limited to the concepts that suffice for modeling the proverbial 80% of practical cases.

2.2 Core Concepts

The core language consists of three main types of elements (note, however, that the model elements often represent *classes* of entities in the real world): *active structure* elements, *behavior* elements, and *passive structure* elements (*objects*). The active structure elements are the business actors, application components, and devices that display actual behavior; i.e., the 'subjects' of activity (right side of the Figure 2).

> An active structure element is defined as an entity that is capable of performing behavior.

Then there is the behavioral or dynamic aspect (center of Figure 2). The active structure concepts are assigned to behavioral concepts, to show who or what performs the behavior.

> A behavior element is defined as a unit of activity performed by one or more active structure elements.

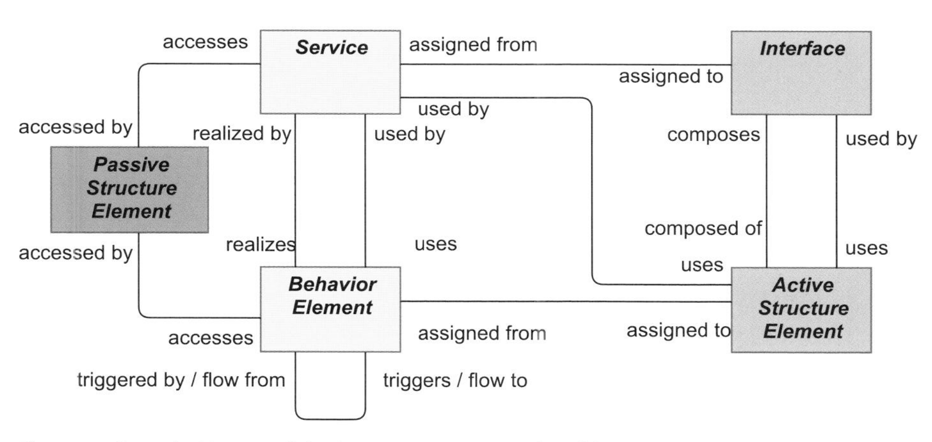

Figure 2: Generic Metamodel: The Core Concepts of ArchiMate[1]

The passive structure elements are the objects on which behavior is performed.

1 In this figure, and all the other metamodel pictures in this document, a convention for role names of relationships is used that is similar to UML (but using verbs instead of nouns). For example, a Behavior Element *realizes* a Service, and a Service *is realized by* a Behavior Element. If no cardinality is shown for a relationship end, a default of 0..* (zero or more) is assumed; if the default does not apply, the cardinality is shown explicitly in the metamodel.

> A passive structure element is defined as an object on which behavior is performed.

In the domain of information-intensive organizations, which is the main focus of the language, passive structure elements are usually information or data objects, but they may also be used to represent physical objects. These three aspects – active structure, behavior, and passive structure – have been inspired by natural language, where a sentence has a subject (active structure), a verb (behavior), and an object (passive structure).

Second, we make a distinction between an external view and an internal view on systems. When looking at the behavioral aspect, these views reflect the principles of service orientation.

> A service is defined as a unit of functionality that a system exposes to its environment, while hiding internal operations, which provides a certain value (monetary or otherwise).

Thus, the service is the externally visible behavior of the providing system, from the perspective of systems that use that service; the environment consists of everything outside this providing system. The value provides the motivation for the service's existence. For the external users, only this exposed functionality and value, together with non-functional aspects such as the quality of service, costs, etc., are relevant. These can be specified in a contract or Service Level Agreement (SLA). Services are accessible through interfaces, which constitute the external view on the active structural aspect.

> An interface is defined as a point of access where one or more services are made available to the environment.

An interface provides an external view on the service provider and hides its internal structure.

2.3 Collaboration and Interaction

Going one level deeper in the structure of the language, we distinguish between behavior that is performed by a *single* structure element (e.g., actor, role component, etc.), or collective behavior (interaction) that is performed by a collaboration of multiple structure elements.

> A collaboration is defined as a (temporary) grouping (or aggregation) of two or more structure elements, working together to perform some collective behavior.

This collective behavior can be modeled as an interaction.

> An interaction is defined as a unit of behavior performed by a collaboration of two or more structure elements.

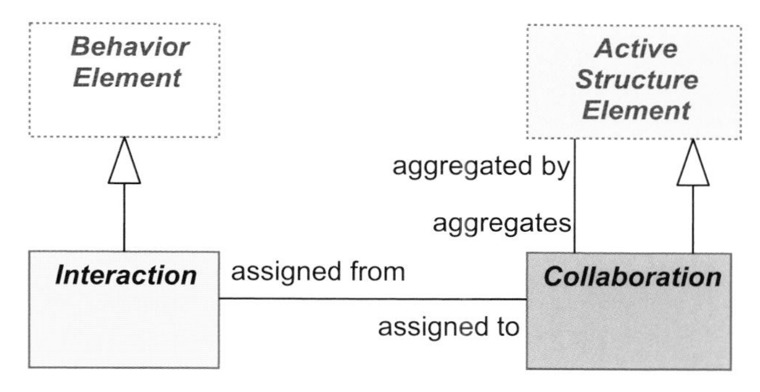

Figure 3: Collaboration and Interaction

2.4 Relationships

Next to the core concepts outlined above, ArchiMate contains a core set of relationships. Several of these relationships have been adopted from corresponding relationship concepts that occur in existing standards; e.g., relationships such as composition, aggregation, association, and specialization are taken from UML 2.0, while triggering is used in many business process modeling languages.

Note: For the sake of readability, the metamodel figures in the next sections do not show all possible relationships in the language. Refer to

Section 6.5 on additional derived relationships. Furthermore, aggregation, composition, and specialization relationships are always permitted between two elements that have the same type.

2.5 Layering

The ArchiMate language defines three main layers (depicted with different colors in the examples in the next chapters), based on specializations of the core concepts described in Sections 2.2 and 2.3:

1. The *Business Layer* offers products and services to external customers, which are realized in the organization by business processes performed by business actors.
2. The *Application Layer* supports the business layer with application services which are realized by (software) applications.
3. The *Technology Layer* offers infrastructure services (e.g., processing, storage, and communication services) needed to run applications, realized by computer and communication hardware and system software.

The general structure of models within the different layers is similar. The same types of concepts and relationships are used, although their exact nature and granularity differ. In Chapters 3, 4, and 5, we further develop these concepts to obtain concepts specific to a particular layer. Figure 2 shows the central structure that is found in each layer.

In line with service orientation, the most important relationship between layers is formed by "used by" relationships, which show how the higher layers make use of the services of lower layers. (Note, however, that services need not only be used by elements in a higher layer, but also can be used by elements in the same layer.) A second type of link is formed by realization relationships: elements in lower layers may realize comparable elements in higher layers; e.g., a "data object" (Application layer) may realize a "business object" (Business layer); or an "artifact" (Technology layer) may realize either a "data object" or an "application component" (Application layer).

2.6 The ArchiMate Framework

The aspects and layers identified in the previous sections can be organized as a framework of nine "cells", as illustrated in Figure 4.

It is important to realize that the classification of concepts based on aspects and layers is only a global one. It is impossible to define a strict boundary between the aspects and layers, because concepts that link the different aspects and layers play a central role in a coherent architectural description. For example, running somewhat ahead of the later conceptual discussions, (business) functions and (business) roles serve as intermediary concepts between "purely behavioral" concepts and "purely structural" concepts.

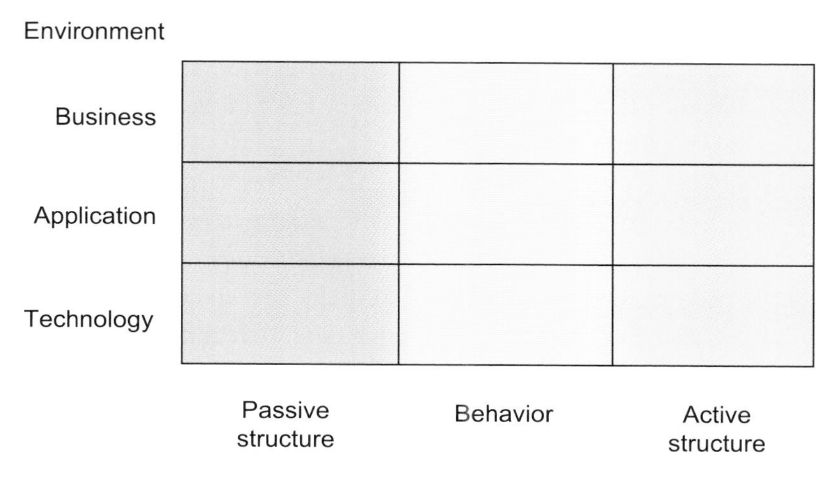

Figure 4: Architectural Framework

Besides the core aspects shown in Figure 4 (passive structure, behavior, and active structure), which are mainly operational in nature, the work of an enterprise architect touches upon numerous other aspects, not explicitly covered by the ArchiMate framework, some of which may cross several (or all) conceptual domains; for example:

- Goals, principles, and requirements
- Risk and security
- Governance
- Policies and business rules
- Costs
- Performance
- Timing
- Planning and evolution

Not all of these aspects can be completely covered using the standard language extension mechanisms as described in Chapter 9. In order to facilitate tool vendors and methodology experts in providing support for these aspects within the overall ArchiMate language, specific extensions can be added. These modular extension add new concepts, relationships, or attributes, while complying to the design restriction that ArchiMate is explicitly designed to be as small as possible.

Also, it may be useful to add concepts or attributes related to the design process rather than to the system or organization that is to be described or designed. Examples of such concepts or attributes are requirements and design decisions.

This new issue of the specification addresses two such extensions: the *Motivation extension* and the *Implementation and Migration extension*. The Motivation extension is introduced in the next section and elaborated in more detail in Chapter 10. The Implementation and Migration extension is introduced in Section 2.8 and elaborated in more detail in Chapter 11. Other aspects may be addressed in future extensions of the language (see Chapter 12 for a more thorough discussion of this).

2.7 Motivation Extension

The core concepts of ArchiMate focus on describing the architecture of systems that support the enterprise. Not covered are the elements which, in different ways, *motivate* the design and operation of the enterprise. These motivational aspects correspond to the "Why" column of the Zachman framework [8], which was intentionally left out of scope in the design of ArchiMate 1.0.

The Motivation extension of ArchiMate adds the motivational concepts such as goal, principle, and requirement. It addresses the way the enterprise architecture is aligned to its context, as described by motivational elements.

> A motivational element is defined as an element that provides the context or reason lying behind the architecture of an enterprise.

In addition, the Motivation extension recognizes the concepts of stakeholders, drivers, and assessments. Stakeholders represent (groups of) persons or organizations that influence, guide, or constrain the enterprise. Drivers represent internal or external factors which influence the plans and aims of an enterprise. An understanding of strengths, weaknesses, opportunities, and threats in relation to these drivers will help the formation of plans and aims to appropriately address these issues.

Figure 5 depicts that the core elements of an architectural description are related to motivational elements via requirements. Goals and principles have to be translated into requirements before core elements, such as services, processes, and applications, can be assigned that realize them. The possible relationships among motivational elements are explained in Chapter 10.

Another relationship between the core metamodel and the Motivation extension is that a business actor may be assigned to a stakeholder, which can be seen as a motivational role (as opposed to an operational business role) that an actor may fulfill.

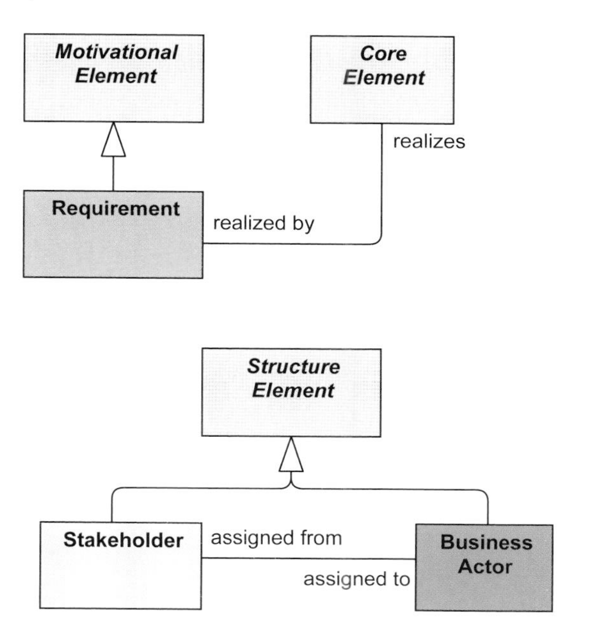

Figure 5: Relationship between Core and Motivational Elements in ArchiMate

The main reason to introduce motivational concepts in ArchiMate is to support requirements management and to support the Preliminary Phase and Phase A (Architecture Vision) of the TOGAF ADM, which establish the high-level business goals, architecture principles, and initial business requirements.

Requirements management is an important activity in the process of designing and managing enterprise architectures. Goals from various stakeholders form the basis for any change to an organization. These goals need to be translated into requirements on the organization's architecture. This architecture should reflect how the requirements are realized by services, processes, and software applications in the day-to-day operations. Therefore, the quality of the architecture is largely determined by the ability to capture and analyze the relevant goals and requirements, the extent to which they can be realized by the architecture, and the ease with which goal and requirements can be changed.

Principles and requirements are strongly related [3]. Principles are general rules and guidelines that help inform and support the way in which an organization sets about fulfilling its mission. In contrast, requirements constrain and shape a specific design of some enterprise architecture. This corresponds to the distinction between two commonly used interpretations of enterprise architecture: (i) as the structure of some organization in terms of its components and their relationships, and (ii) as a set of principles that should be applied to any such structure.[2] The scope of the first interpretation concerns a single design of the organization, whereas the second concerns any possible design. Requirements are associated with the first interpretation. Instead, principles are independent of a specific design and have to be specialized into requirements in the process of designing the organization's architecture. This makes the application of principles an important part of requirements management.

Inadequate requirements management is one of the main causes of impaired or failed IT projects [21], due to exceeding budgets or deadlines, or not delivering the expected results. This is well phrased by the following quote of Brooks [22]: "No other part of the work so cripples the resulting system

2 Both interpretations are combined in the second meaning of architecture as described in Section 1.2.

if done wrong". Therefore, the requirements management process and the architecture development process need to be well-aligned, and traceability should be maintained between requirements and the architectural elements that realize these requirements.

In the TOGAF Architecture Development Method (ADM) [4], requirements management is a central process that applies to all phases of the ADM cycle. While TOGAF presents "requirements" on requirements management, it refrains from mandating or recommending existing languages, methods, and tools from the area of requirements engineering. ArchiMate supports the requirements management process by means of the motivational concepts.

2.8 Implementation and Migration Extension

The Implementation and Migration extension of ArchiMate adds concepts to support the late ADM phases, related to the implementation and migration of architectures: Phase E (Opportunities and Solutions), Phase F (Migration Planning), and Phase G (Implementation Governance).

This extension includes concepts for modeling implementation programs and projects to support program, portfolio, and project management, and a plateau concept to support migration planning. The proposed extension aims at covering the main concepts of program and project management standards and best practices, such as MSP [23], PRINCE2 [24], and PMBoK [25]. Concepts that are specific to one of these methods are not part of the extension, but may be defined as specialization of the generic concepts. In this way, the set of concepts and relationships that are defined in the extension is kept at a minimum.

Furthermore, concepts or relationships from the ArchiMate core or the Motivation extension are re-used where possible. Figure 6 depicts the relationship between concepts from the Implementation and Migration extension and concepts from the ArchiMate core and Motivation extension. A deliverable may realize core elements within an architecture. A gap may be associated with any number of core elements. A location may be assigned to work packages and deliverables. A work package realizes requirements indirectly through the realization of core elements (e.g., an application component, business process, or service). Also, core elements are linked to the other concepts of the Motivation extension by means of derived relationships. The possible relationships among implementation and migration, core, and motivational elements are explained in more detail in Chapters 10 and 11.

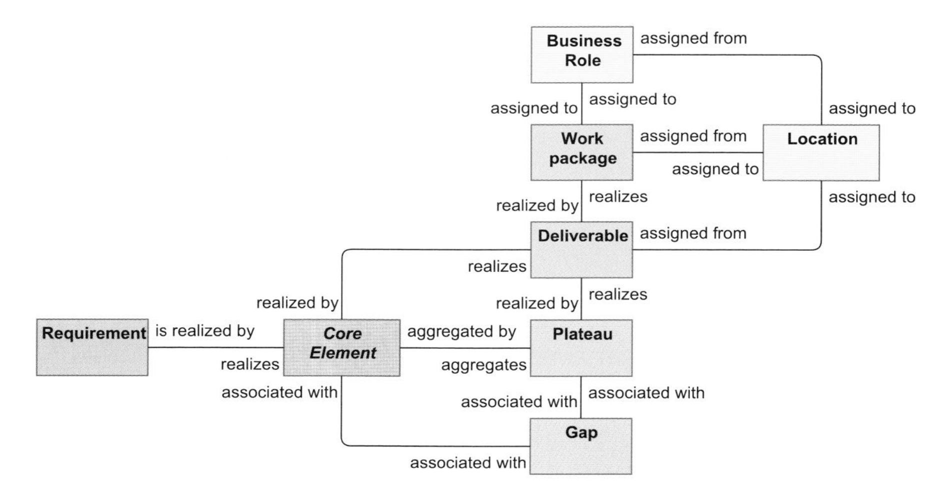

Figure 6: Relationships between Motivational, Core, and Implementation and Migration Elements

2.9 ArchiMate and TOGAF

The ArchiMate language, as described in this Technical Standard, complements TOGAF [4] in that it provides a vendor-independent set of concepts, including a graphical representation, that helps to create a consistent, integrated model "below the waterline", which can be depicted in the form of TOGAF views.

The structure of the core ArchiMate language closely corresponds with the three main architectures as addressed in the TOGAF ADM. This is illustrated in Figure 7. This correspondence would suggest a fairly easy mapping between TOGAF views and the ArchiMate viewpoints.

Some TOGAF views are not matched in the ArchiMate core, however. Partially, this is because the scope of TOGAF is broader and in particular addresses more of the high-level strategic issues and the lower-level engineering aspects of system development, whereas the ArchiMate core is limited to the enterprise architecture level of abstraction. However,

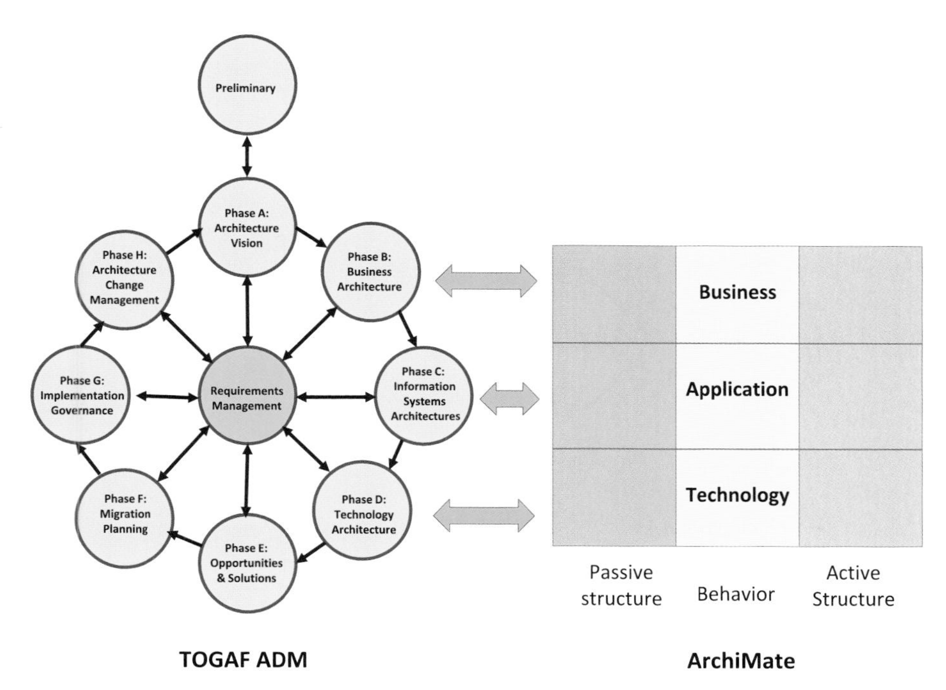

Figure 7: Correspondence between ArchiMate and TOGAF

the two language extensions, described in Chapters 10 and 11, address these additional issues. They define concepts such as goal, principle, and requirement, as well as the planning and migration-oriented concepts. Figure 8 illustrates this.

Although some of the viewpoints that are defined in TOGAF cannot easily be mapped onto ArchiMate viewpoints, the ArchiMate language and its analysis techniques do support the concepts addressed in these viewpoints. While there is no one-to-one mapping between them, there is still a fair amount of correspondence between the ArchiMate viewpoints and the viewpoints that are defined in TOGAF. Although corresponding viewpoints from ArchiMate and TOGAF do not necessarily have identical coverage, we can see that many viewpoints from both methods address largely the same issues.

TOGAF and ArchiMate can easily be used in conjunction and they appear to cover much of the same ground, although with some differences in scope and approach.

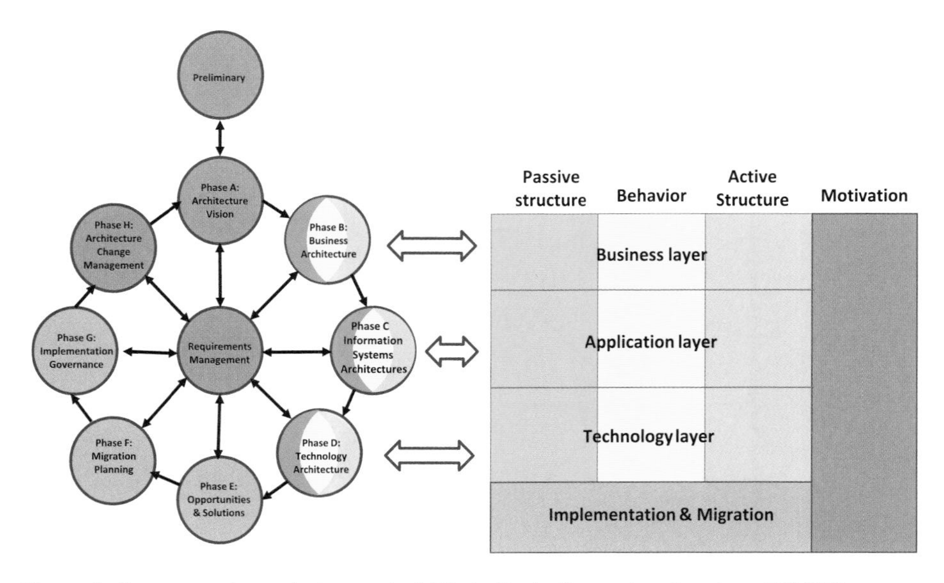

Figure 8: Correspondence between ArchiMate (including extensions) and TOGAF

Chapter 3

Business Layer

3.1 Business Layer Metamodel

Figure 9 shows the metamodel of business layer concepts. The metamodel follows the structure of the generic metamodel introduced in the previous chapter. However, this layer also includes a number of additional informational concepts which are relevant in the business domain: a product and associated contract, the meaning of business objects, and the value of products and business services.

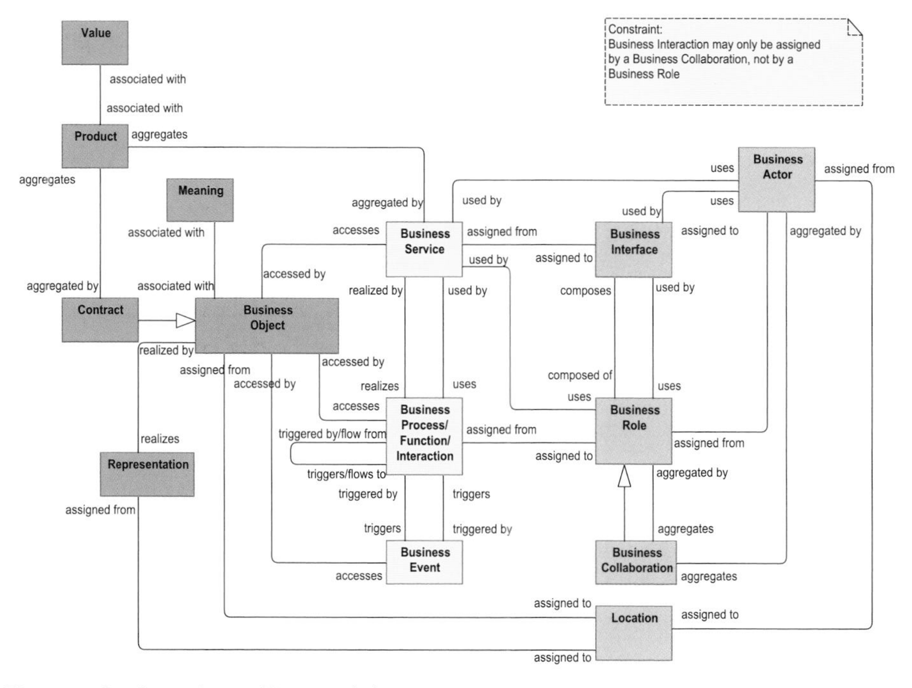

Figure 9: Business Layer Metamodel[1]

1 In the metamodel pictures, we use colors to distinguish concepts belonging to the different aspects of the ArchiMate framework: green for passive structure, yellow for behavior, and blue for active structure. In ArchiMate models, there are no formal semantics assigned to colors. However, they can be used freely to stress certain aspects in models. For instance, in the example models presented in this standard, we often use colors to distinguish between the layers of the ArchiMate framework: yellow for the business layer, blue for the application layer, and green for the technology layer.

Note: This figure does not show all permitted relationships: every concept in the language can have composition, aggregation, and specialization relationships with concepts of the same type; furthermore, there are indirect relationships that can be derived, as explained in Section 6.5.

3.2 Structural Concepts

The structure aspect at the business layer refers to the static structure of an organization, in terms of the entities that make up the organization and their relationships.

Two types of entities are distinguished:

- The *active entities* that are the subjects (e.g., business actors or business roles) that perform behavior such as business processes or functions (capabilities). Business actors may be individual persons (e.g., customers or employees), but also groups of people (organization units) and resources that have a permanent (or at least long-term) status within the organizations. Typical examples of the latter are a department and a business unit.
- The *passive entities* (business objects) that are manipulated by behavior such as business processes or functions. The passive entities represent the important concepts in which the business thinks about a domain.

Architectural descriptions focus on structure, which means that the inter-relationships of entities within an organization play an important role. To make this explicit, the concept of business collaboration has been introduced. Business collaborations have been inspired by collaborations as defined in the UML 2.0 standard [7], [10], although the UML collaborations apply to components in the application layer. Also, the ArchiMate business collaboration concept has a strong resemblance to the "community" concept as defined in the RM-ODP Enterprise Language [6], as well as to the "interaction point" concept, defined in Amber [11] as the place where interactions occur.

The concept of business interfaces is introduced to explicitly model the (logical or physical) locations or channels where the services that a role offers to the environment can be accessed. The same service may be offered on a number of different interfaces; e.g., by mail, by telephone, or through the Internet. In contrast to application modeling, it is uncommon in current business layer modeling approaches to recognize the business interface concept.

3.2.1 Business Actor

A business actor is defined as an organizational entity that is capable of performing behavior.

A business actor performs the behavior assigned to (one or more) business roles. A business actor is an organizational entity as opposed to a technical entity; i.e., it belongs to the business layer. Actors may, however, include entities outside the actual enterprise; e.g., customers and partners. Examples of business actors are humans, departments, and business units. A business actor may be assigned to one or more business roles. The name of a business actor should preferably be a noun.

Figure 10: Business Actor Notation

Example

The model below illustrates the use of business actors. The company ArchiSurance is modeled as a business actor that is composed of two departments. The **Travel insurance seller** role is assigned to the travel department. In this role, the travel department performs the **Take out insurance** process, which offers a service that is accessible via the business interface assigned to this role.

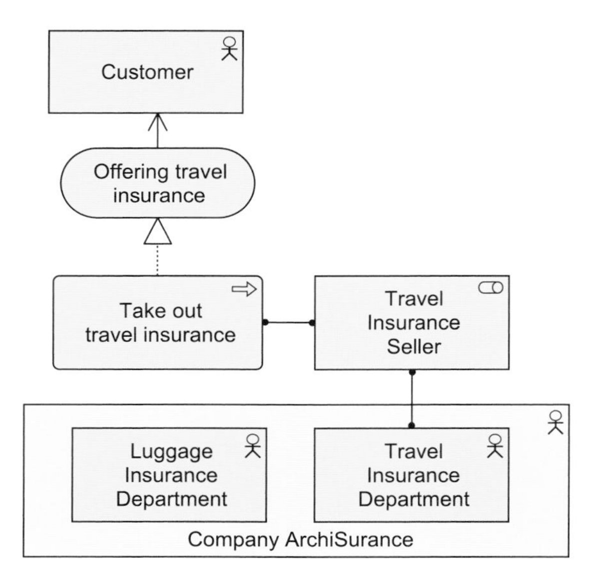

Example 1: Business Actor

3.2.2 Business Role

> A business role is defined as the responsibility for performing specific behavior, to which an actor can be assigned.

Business processes or business functions are assigned to a single business role with certain responsibilities or skills. A business actor that is assigned to a business role ultimately performs the corresponding behavior. In addition to the relation of a business role with behavior, a business role is also useful in a (structural) organizational sense; for instance, in the division of labor within an organization.

A business role may be assigned to one or more business processes or business functions, while a business actor may be assigned to a business role. A business interface or an application interface may be used by a business role, while a business interface may be part of a business role (through a composition relationship, which is not shown explicitly in the interface notation). The name of a business role should preferably be a noun.

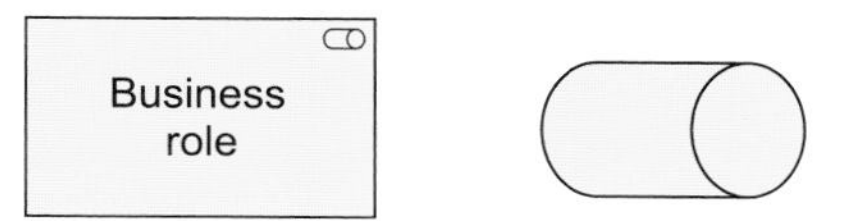

Figure 11: Business Role Notation

Example

In the model below, the business role Insurance Seller is fulfilled by the Insurance Department actor and has telephone as a provided interface. The business role Insurance Buyer is fulfilled by the Customer actor, and has telephone as a required interface.

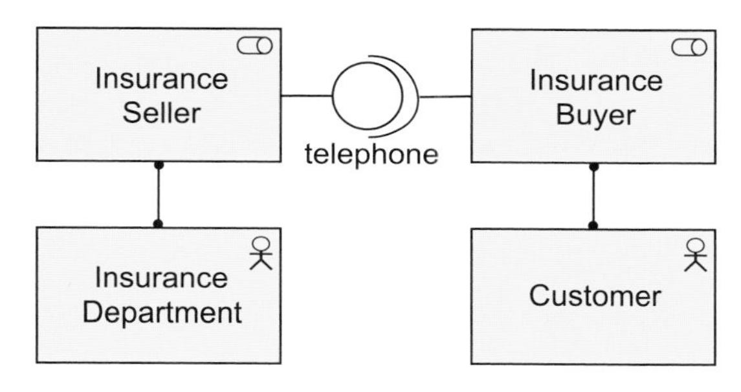

Example 2: Business Role

3.2.3 Business Collaboration

Business collaboration is defined as an aggregate of two or more business roles that work together to perform collective behavior.

A business process or function may be interpreted as the internal behavior assigned to a single business role. In some cases behavior is the collective effort of more than one business role; in fact a collaboration of two or more business roles results in collective behavior which may be more than simply the sum of the behavior of the separate roles. Business collaborations represent this collective effort. Business interactions are used to describe the internal behavior that takes place within business collaboration.
A collaboration is a (possibly temporary) collection of roles within an

organization which perform collaborative behavior (interactions). Unlike a department, which may also group roles, a business collaboration does not have an official (permanent) status within the organization; it is specifically aimed at a specific interaction or set of interactions between roles. However, a business collaboration can be regarded as a kind of "virtual role", hence its designation as a specialization of role. It is especially useful in modeling B2B interactions between different organizations.

A business collaboration may be composed of a number of business roles, and may be assigned to one or more business interactions. A business interface or an application interface may be used by a business collaboration, while a business collaboration may have business interfaces (through composition). The name of a business collaboration should preferably be a noun. It is also rather common to leave a business collaboration unnamed.

Figure 12: Business Collaboration Notation

> **Example**
> The model below illustrates a possible use of the collaboration concept. In this example, selling an insurance product involves the **Sales department**, fulfilling a sales support role, and a department specialized in that particular type of insurance, fulfilling an insurance seller role. The example also shows that one role, in this case **Sales support**, can participate in more than one collaboration.

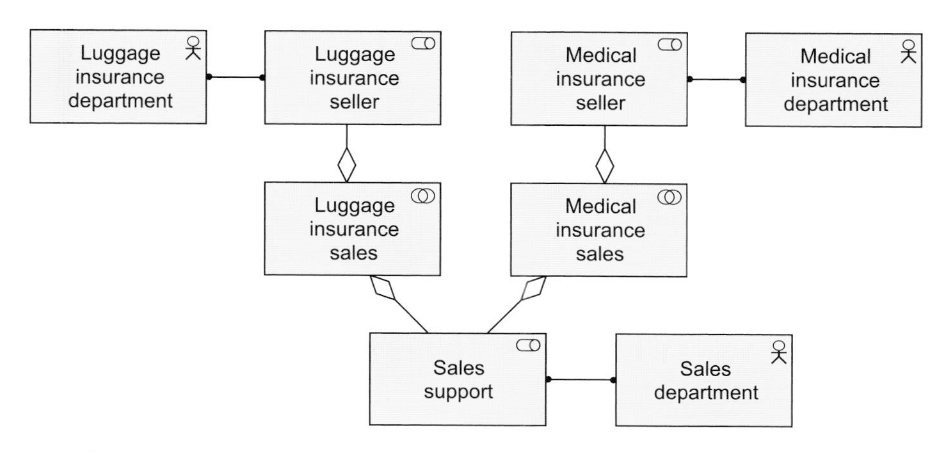

Example 3: Business Collaboration

3.2.4 Business Interface

> A business interface is defined as a point of access where a business service is made available to the environment.

A business interface exposes the functionality of a business service to other business roles (provided interface), or expects functionality from other business services (required interface). It is often referred to as a channel (telephone, internet, local office, etc.). The same business service may be exposed through different interfaces.

A business interface may be part of a business role through a composition relationship, which is not shown in the standard notation, and a business interface may be used by a business role. A business interface may be assigned to one or more business services, which means that these services are exposed by the interface. The name of a business interface should preferably be a noun.

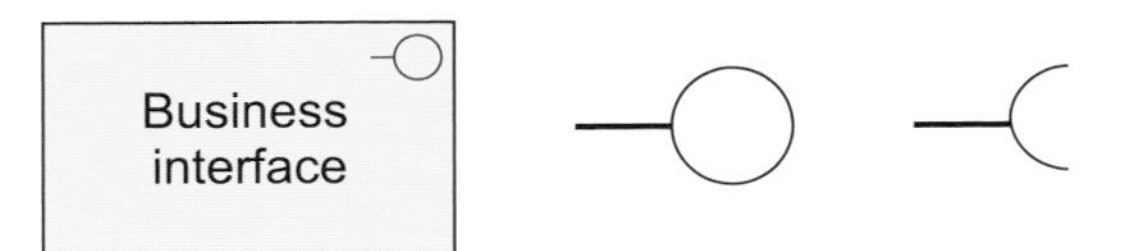

Figure 13: Business Interface Notation

Example
In the model below, the business services provided by the **Luggage insurance seller** and its collaboration with the **Medical insurance seller** are exposed by means of a web form and call center business interface, respectively.

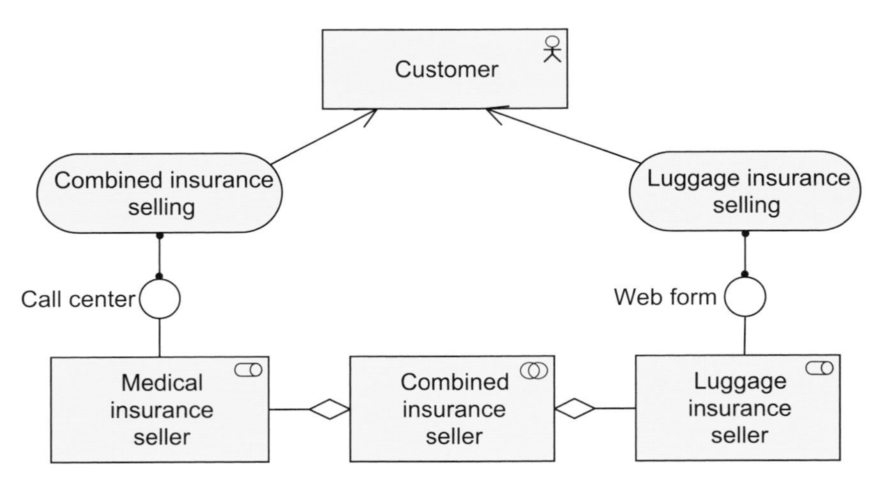

Example 4: Business Interface

3.2.5 Location

A location is defined as a conceptual point or extent in space.

The location concept is used to model the distribution of structural elements such as business actors, application components, and devices. This is modeled by means of an assignment relationship from location to structural element. Indirectly, a location can also be assigned to a behavior element, to indicate where the behavior is performed.

Figure 14: Location Notation

Example

The model below shows that the departments of an insurance company are distributed over different locations. The Legal and Finance departments are centralized at the main office, and there are claims handling departments at various local offices throughout the country.

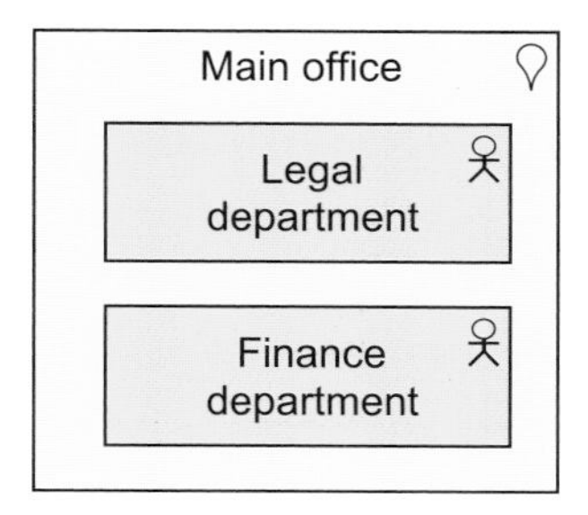

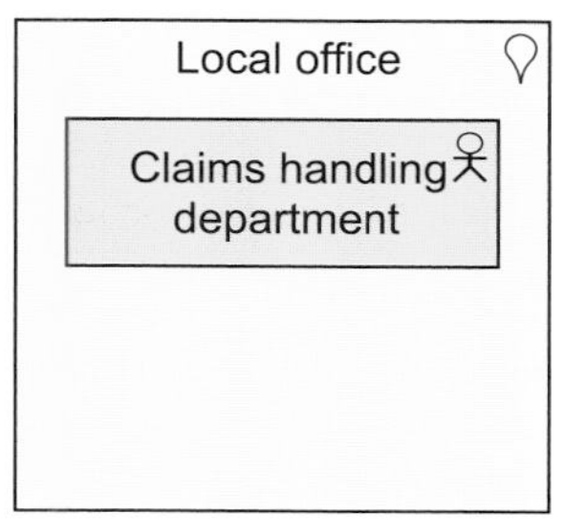

Example 5: Location

3.2.6 Business Object

A business object is defined as a passive element that has relevance from a business perspective.

Business objects represent the important "informational" or "conceptual" elements in which the business thinks about a domain. Generally, a business object is used to model an object type (cf. a UML class), of which several instances may exist within the organization. A wide variety of types of business objects can be defined. Business objects are passive in the sense that they do not trigger or perform processes.

Business objects may be accessed (e.g., in the case of information objects, which are most common in the application domains in which ArchiMate is applied, they may be created, read, written) by a business process, function, a business interaction, a business event, or a business service. A business object may have association, specialization, aggregation, or composition relationships with other business objects. A business object may be realized by a representation or by a data object (or both). The name of a business object should preferably be a noun.

Business
object

Figure 15: Business Object Notation

Example

The model below shows a business object **Invoice,** which aggregates (multiple) business objects **Invoice line**. Two possible realizations of this business object exist: an **Electronic invoice** (data object) and a **Paper invoice** (representation). The business process **Create invoice** creates the invoice and the invoice lines, while the business process **Send invoice** accesses the business object **Invoice.**

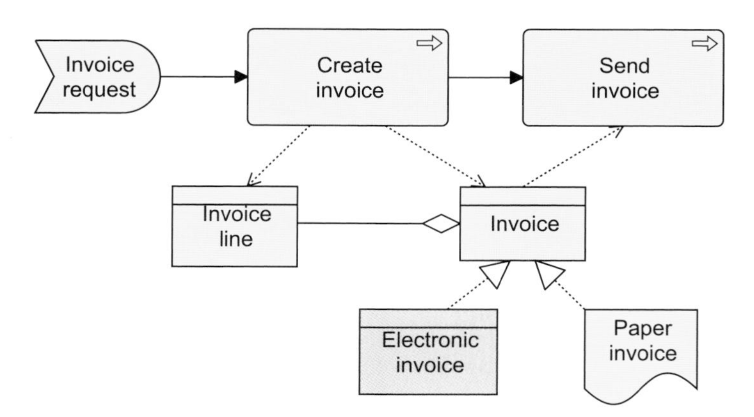

Example 6: Business Object

3.3 Behavioral Concepts

Based on service orientation, a crucial design decision for the behavioral part of our metamodel is the distinction between "external" and "internal" behavior of an organization.

The externally visible behavior is modeled by the concept *business service*. A business service represents a coherent piece of functionality that offers added value to the environment, independent of the way this functionality is realized internally. A distinction can be made between "external" business services, offered to external customers, and "internal" business services, offering supporting functionality to processes or functions within the organization.

Several types of internal behavior elements that can realize a service are distinguished. Although the distinction between the two is not always sharp, it is often useful to distinguish a *process view* and a *function view* on behavior; two concepts associated with these views, *business process* and *business function*, are defined. Both concepts can be used to group more detailed business processes/functions, but based on different grouping criteria. A *business process* represents a workflow or value stream consisting of smaller processes/functions, with one or more clear starting points and leading to some result. It is sometimes described as "customer to customer", where this customer may also be an internal customer, in the case of sub-processes within an organization. The goal of such a business process is to "satisfy or delight the customer" [12]. A *business function* offers functionality that may be useful for one or more business processes. It groups behavior based on, for example, required skills, resources, (application) support, etc. Typically, the business processes of an organization are defined based on the *products* and *services* that the organization offers, while the business functions are the basis for, for example, the assignment of resources to tasks and the application support.

A *business interaction* is a unit of behavior similar to a business process or function, but which is performed in a collaboration of two or more roles within the organization. Unlike the interaction concept in Amber [11], which is an *atomic* unit of collaborative behavior, our business interaction can be decomposed into smaller interactions. Although interactions are external behavior from the perspective of the roles participating in the collaboration, the behavior is internal to the collaboration as a whole. Similar to processes or functions, the result of a business interaction can be made available to the environment through a business service.

A *business event* is something that happens (externally) and may influence business processes, functions, or interactions. The "business event" concept is similar to the "trigger" concept in Amber [11] and the "initial state" and "final state" concepts as used in, for example, UML activity diagrams. However, our business event is more generally applicable in the sense that it can also be used to model other types of events, in addition to triggers.

3.3.1 Business Process

> A business process is defined as a behavior element that groups behavior based on an ordering of activities. It is intended to produce a defined set of products or business services.

A business process describes the internal behavior performed by a business role that is required to produce a set of products and services. For a consumer the products and services are relevant and the required behavior is merely a black box, hence the designation "internal".

In comparison to a business interaction, in which a collaboration of two or more business roles are (interactively) involved, at a given level of granularity only one business role is involved with a business process. However, a complex business process may be an aggregation of other, finer-grained processes, each of which may be assigned to finer-grained roles that are aggregated by roles that are aggregated by the original role.

There is a potential many-to-many relationship between business processes and business functions. Informally speaking, processes describe some kind of "flow" of activities, whereas functions group activities according to required skills, knowledge, resources, etc.

A business process may be triggered by, or trigger, any other business behavior element (e.g., business event, business process, business function, or business interaction). A business process may access business objects.

A business process may realize one or more business services and may use (internal) business services or application services. A business role or an application component may be assigned to a business process to perform this process manually or automated, respectively. The name of a business process should preferably be a verb in the simple present tense; e.g., "handle claim".

In an ArchiMate model, the existence of business processes is depicted. It does not, however, list the flow of activities in detail. During business process modeling, a business process can be expanded using a business process design language; e.g., BPMN [20].

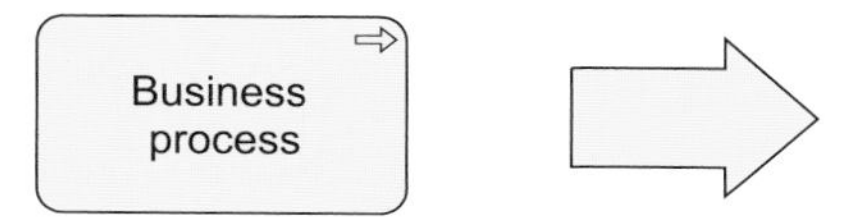

Figure 16: Business Process Notation

Example

The model below illustrates the use of business processes and its relation with other concepts. The **Take out insurance** process is composed of three sub-processes. For clarity, the sub-processes are drawn in the overall process (structuring). Each sub-process triggers the next sub-process. The event **Request for Insurance** triggers the first sub-process. A particular role, in this case an insurance seller, is assigned to perform the required work. The process itself realizes an **Insurance selling** service. The **Receive request** sub-process uses the business object **Customer info.**

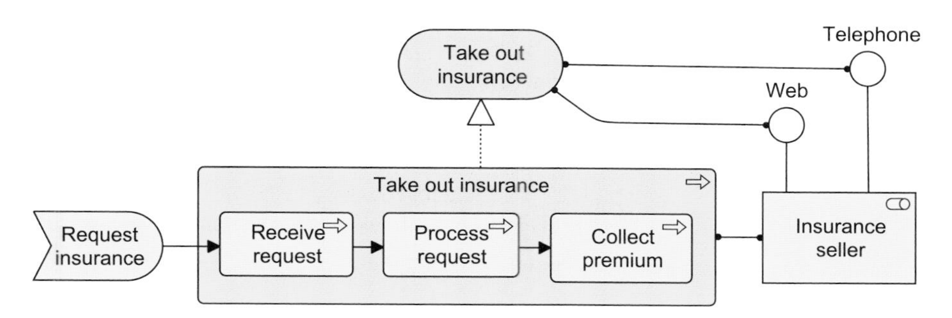

Example 7: Business Process

3.3.2 Business Function

A business function is defined as a behavior element that groups behavior based on a chosen set of criteria (typically required business resources and/or competences).

Just like a business process, a business function also describes internal behavior performed by a business role. However, while a business process group's behavior is based on a sequence or "flow" of activities that is needed to realize a product or service, a business function typically groups behavior based on required business resources, skills, competences, knowledge, etc.

There is a potential many-to-many relation between business processes and business functions. Complex processes in general involve activities that offer various functions. In this sense a business process forms a string of business functions. In general, a business function delivers added value from a business point of view. Organizational units or applications may coincide with business functions due to their specific grouping of business activities.

A business function may be triggered by, or trigger, any other business behavior element (business event, business process, business function, or business interaction). A business function may access business objects. A business function may realize one or more business services and may use (internal) business services or application services. A business role or an application component may be assigned to a business function. The name of a business function should preferably be a verb ending with "-ing"; e.g., "claims processing", or a noun ending in "-ion" or "-ment"; e.g., "administration".

Figure 17: Business Function Notation

Example

The model below illustrates the use of business functions, as well as the relationship between business functions and business processes. The three business functions group a number of business sub-processes. The business process, initiated by a business event, involves sub-processes from different business functions. The Insurer role is assigned to each of the business functions. Moreover, business functions may access business objects; in this case, the **Customer handling** function uses or manipulates the **Customer information** object. Also, the **Financial handling** function makes use of a **Billing** application service and realizes a **Premium collection** business service.

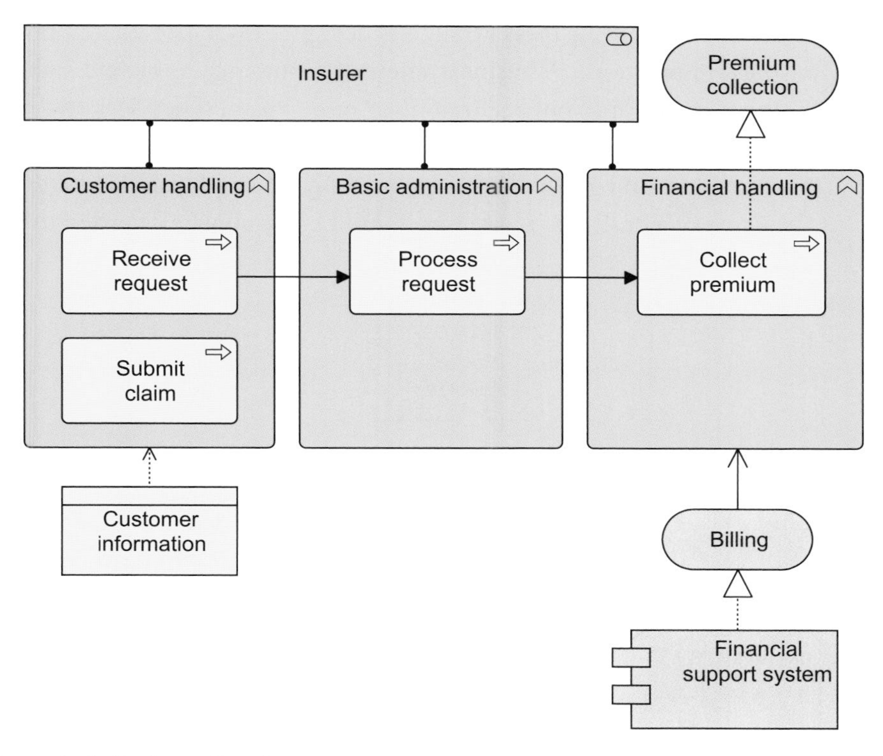

Example 8: Business Function

3.3.3 Business Interaction

> A business interaction is defined as a behavior element that describes the behavior of a business collaboration.

A business interaction is similar to a business process/function, but while a process/function may be performed by a single role, an interaction is performed by a collaboration of multiple roles. The roles in the collaboration share the responsibility for performing the interaction.

A business interaction may be triggered by, or trigger, any other business behavior element (business event, business process, business function, or business interaction). A business interaction may access business objects. A business interaction may realize one or more business services and may use (internal) business services or application services. A business collaboration or an application collaboration may be assigned to a business interaction. The name of a business interaction should preferably be a verb in the simple present tense.

Figure 18: Business Interaction Notation

> **Example**
> In the model below, a business interaction is triggered by a request. The business interaction **Take out combined insurance** is performed as collaboration between the travel and luggage insurance seller. The business interaction needs the **Policy info** business object, and realizes the (external) business service **Combined insurance selling**. As part of the business interaction, the **Prepare travel policy** and **Prepare luggage policy** are triggered. The **Travel insurance seller** and **Luggage insurance seller** perform these processes separately.

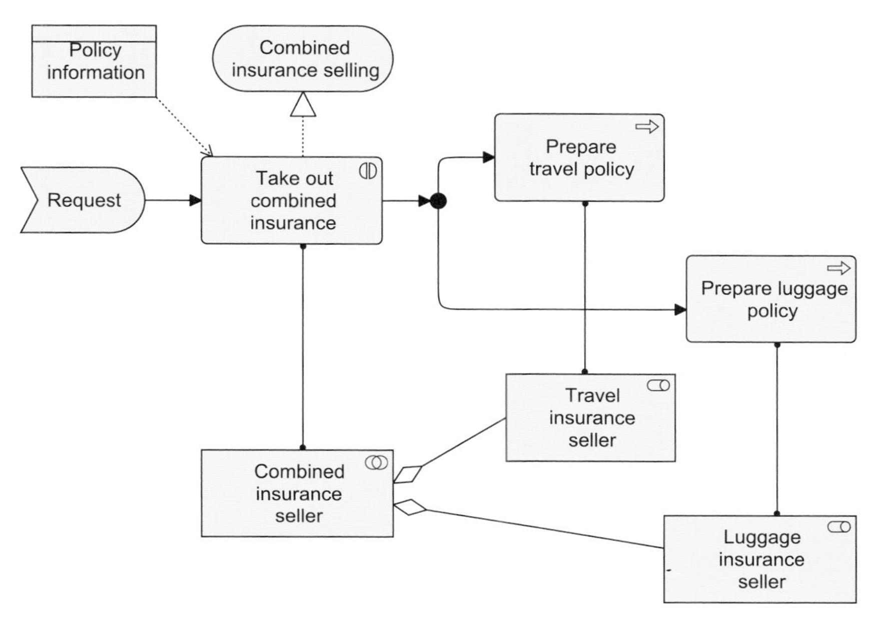

Example 9: Business Interaction

3.3.4 Business Event

> A business event is defined as something that happens (internally or externally) and influences behavior.

Business processes and other business behavior may be triggered or interrupted by a business event. Also, business processes may raise events that trigger other business processes, functions, or interactions. A business event is most commonly used to model something that triggers behavior, but other types of events are also conceivable; e.g., an event that interrupts a process. Unlike business processes, functions, and interactions, a business event is instantaneous: it does not have duration. Events may originate from the environment of the organization (e.g., from a customer), but also internal events may occur generated by, for example, other processes within the organization.

A business event may trigger or be triggered (raised) by a business process, business function, or business interaction. A business event may access a business object and may be composed of other business events. The name of a business event should preferably be a verb in the perfect tense; e.g., "claim received".

Figure 19: Business Event Notation

> **Example**
>
> In the model below, the **Request insurance** event triggers the **Take out insurance** process. A business object containing the **Customer info** accompanies the request. In order to persuade the customer to purchase more insurance products, a triggering event is raised in the **Receive request** process. This triggers the **Send product portfolio to customer** process.

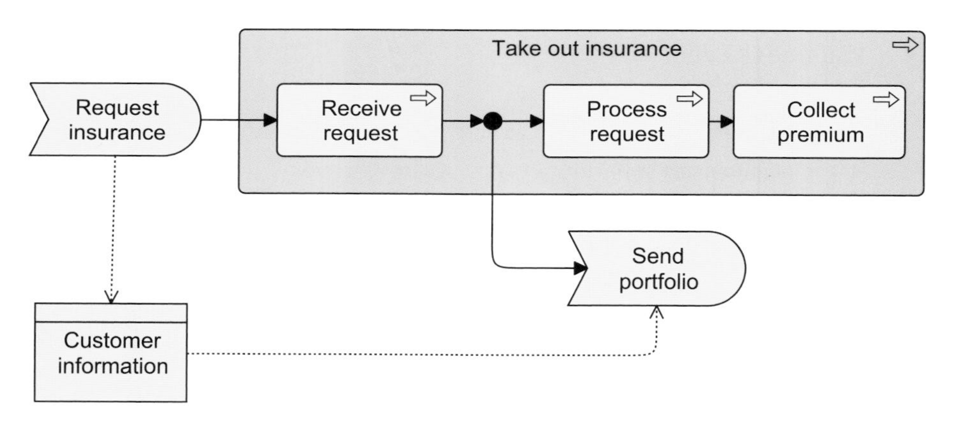

Example 10: Business Event

3.3.5 Business Service

> A business service is defined as a service that fulfills a business need for a customer (internal or external to the organization).

A business service exposes the functionality of business roles or collaborations to their environment. This functionality is accessed through one or more business interfaces. A business service is realized by one or more business processes, business functions, or business interactions that are performed by the business roles or business collaborations, respectively. It may access business objects.

A business service should provide a unit of functionality that is meaningful from the point of view of the environment. It has a purpose, which states this utility. The environment includes the (behavior of) users from outside as well as inside the organization. Business services can be external, customer-facing services (e.g., a travel insurance service) or internal support services (e.g., a resource management service).

A business service is associated with a value. A business service may be used by a business process, business function, or business interaction. A business process, business function, or business interaction may realize a business service. A business interface or application interface may be assigned to a business service. A business service may access business objects. The name of a business service should preferably be a verb ending with "-ing"; e.g., "transaction processing". Also, a name explicitly containing the word "service" may be used.

Figure 20: Business Service Notation

Example

In the model below, external and internal business services are distinguished. The **Basic administration** function acts as a shared service center. The take out business processes corresponding with the travel and luggage insurance use the (internal) business services that are provided by the **Basic administration** function. Both business processes realize an (external) business service. The insurance selling service is accessible via a business interface (e.g., web form) of the insurance seller. Each business service should be of value to the user(s) of the service (in this example, the insurance buyer role). This value may be explicitly modeled, if appropriate. The value of the **Travel insurance selling** service to an external customer (the insurance buyer) is that the customer is insured.

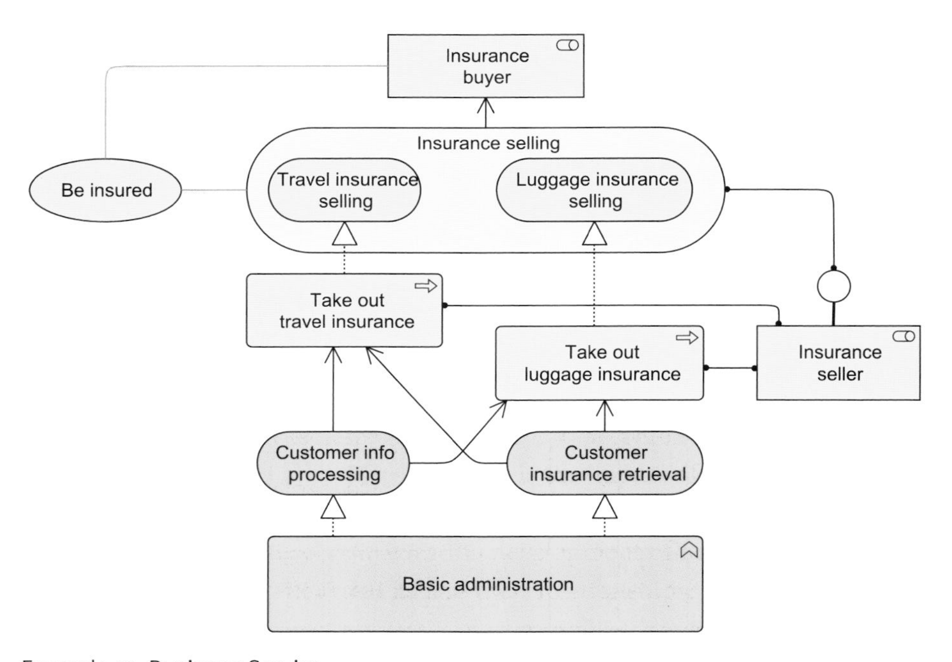

Example 11: Business Service

3.4 Informational Concepts

In contrast to the structural and behavioral concepts, which are mainly concerned with the operational perspective on an enterprise, the informational concepts focus on what we could call the "intentional" perspective. They provide a way to link the operational side of an organization to the business goals, and to the products that an organization offers to its customers. We also classify the product concept itself, together with the related contract concept, as informational concepts.

Information is fundamentally related to communication. Information always serves a particular purpose, which is tightly connected to some communicational goal. As communication always involves a static part (the "message") and a dynamic part (the communication action itself), the communicational goals may have a link to both our "meaning" concept and our "value" concept. Also, in speech act-based approaches to business modeling, such as DEMO [9], the communicational aspect plays a central role in the context of business transactions.

A *representation* is the perceptible form of the information carried by a business object, such as a document. As such, it can be seen as the realization of the associated business object. If relevant, representations can be classified in various ways; for example, in terms of medium (e.g., electronic, paper, audio) or format (e.g., HTML, PDF, plain text, bar chart).

A *meaning* is the contribution of a business object or its representation to the knowledge or expertise of some actor, given a particular context (e.g., the role that the actor fulfills within that context). In other words, meaning represents the informative value of a business object for a user of such an object. It is through a certain interpretation of a representation of the object that meaning is being offered to a certain user or to a certain category of users. A meaning can very well be a reformulation or transformation of parts of the object representation in such a way that the role of the meaning is immediately clear within the user's world, but essentially lies in interpretation by individuals, in context.

For the complete description of a meaning, the following two elements are needed, in addition to the representations (and, indirectly, business objects) with which the meaning is associated:

- Some sort of *meaning description*: A meaning description is not equal to the representation causing the meaning: it is a specialized description that aims to clarify or stipulate a meaning. Natural language may be used for this, but also formal languages or diagrams. Typical examples of meaning descriptions are definitions, ontologies, paraphrases, subject descriptions, and tables of content. Meaning descriptions may draw from or refer to additional meaning description sources; for example, dictionaries. Importantly, meaning descriptions *do not necessarily have to describe meaning in detail*. The level of detail depends on the types of analysis required. It is quite possible that a very rough meaning description is *good enough* to capture at architecture level the sort of interpretations a business object conveys. Detailed meaning description can only in a limited number of cases be made very precise; in most cases, interpretation depends on the general language and knowledge of specific actors, which normally remains largely implicit.

- A description of the *context(s)* in which the meaning is conveyed: A context description covers the situation in which the interpretation takes place. The most important elements of such a context are the *actors sending and receiving the business object*, the *time and place* of communication and the *environment in which this happens*. Often, a context can be briefly described in terms of some business domain.

We see a (financial or information) *product* as of a collection of services, together with a contract that specifies the characteristics, rights, and requirements associated with the product. This "package" is offered as a whole to (internal or external) customers.

We define a *contract* as a formal or informal specification of agreement that specifies the rights and obligations associated with a product. The *value* of a product or service is that which makes some party appreciate it, possibly in relation to providing it, but more typically to acquiring it.

3.4.1 Representation

> A representation is defined as a perceptible form of the information carried by a business object.

Representations (for example, messages or documents) are the perceptible carriers of information that are related to business objects. If relevant, representations can be classified in various ways; for example, in terms of medium (electronic, paper, audio, etc.) or format (HTML, ASCII, PDF, RTF, etc.). A single business object can have a number of different representations. Also, a single representation can realize one or more specific business objects.

A representation may realize one or more business objects. A meaning can be associated with a representation that carries this meaning. The name of a representation is preferably a noun.

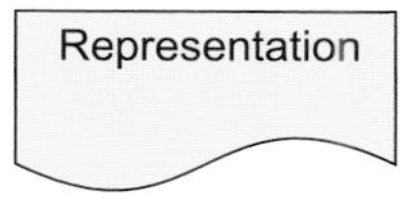

Figure 21: Representation Notation

> **Example**
> The model below shows the business object **Request for insurance**, which is realized (represented) by a (physical) request form. The **Invoice** business object is realized (represented) by a paper bill.

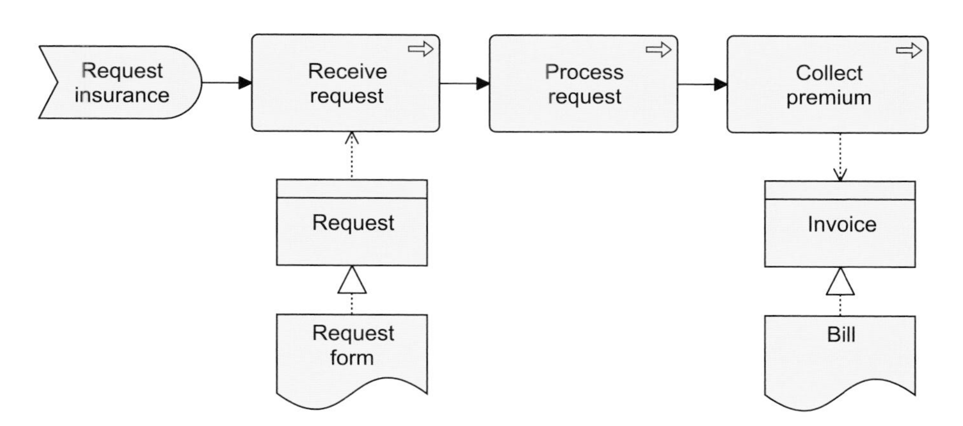

Example 12: Representation

3.4.2 Meaning

> Meaning is defined as the knowledge or expertise present in a business object or its representation, given a particular context.

A meaning is the information-related counterpart of a value: it represents the intention of a business object or representation (for example, a document, message; the representations related to a business object). It is a description that expresses the *intent* of a representation; i.e., how it informs the *external user*.

It is possible that different users view the informative functionality of a business object or representation differently. For example, what may be a "registration confirmation" for a client could be a "client mutation" for a CRM department (assuming for the sake of argument that it is modeled as an external user). Also, various different representations may carry essentially the same meaning. For example, various different documents (a web document, a filled-in paper form, a "client contact" report from the call center) may essentially carry the same meaning.

A meaning can be associated with a representation that carries this meaning. The name of a meaning should preferably be a noun or noun phrase.

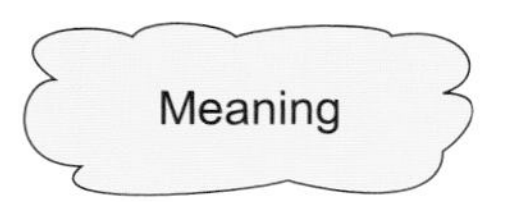

Figure 22: Meaning Notation

> **Example**
> The model below shows an **Insurance policy document** that is the representation of an **Insurance policy,** which is a business object. The meaning related to this document is the **Insurance policy notification**, which consists of a **Policy explanation**, an **Insurance registration,** and a **Coverage description**.

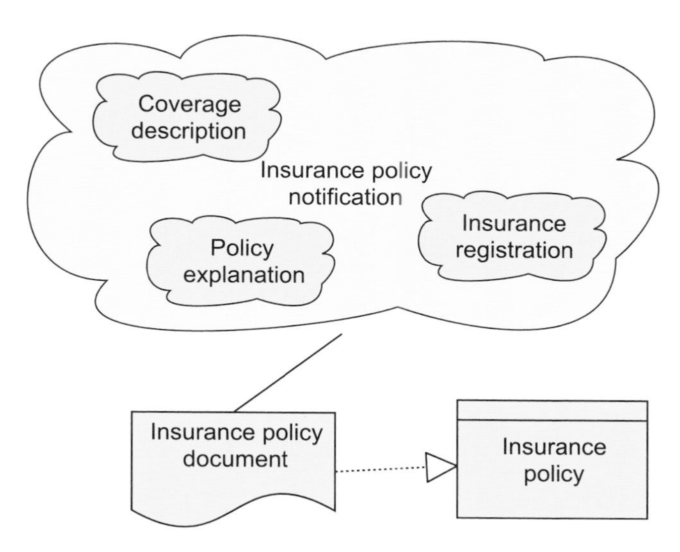

Example 13: Meaning

3.4.3 Value

Value is defined as the relative worth, utility, or importance of a business service or product.

Value may apply to what a party gets by selling or making available some product or service, or it may apply to what a party gets by buying or obtaining access to it. Value is often expressed in terms of money, but it has long since been recognized that non-monetary value is also essential to business; for example, practical/functional value (including the *right* to use a service), and the value of information or knowledge. Though value can hold internally for some system or organizational unit, it is most typically applied to *external* appreciation of goods, services, information, knowledge, or money, normally as part of some sort of customer-provider relationship.

A value can be associated with business services and, indirectly, with the products they are part of, and the roles or actors that use them. Although the name of a value can be expressed in many different ways (including amounts, objects), where the "functional" value of a service is concerned it is recommended to try and express it as an action or state that can be performed or reached as a result of the corresponding service being available.

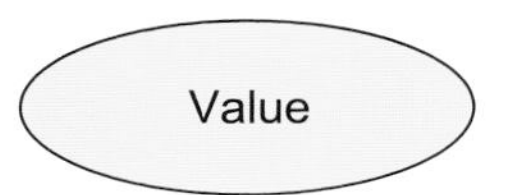

Figure 23: Value Notation

Example
In the model below, the value **Be Insured** is the highest-level expression of what the service **Provide Insurance** enables the client to do; three "sub-values" are distinguished that are part of what **Be Insured** amounts to.

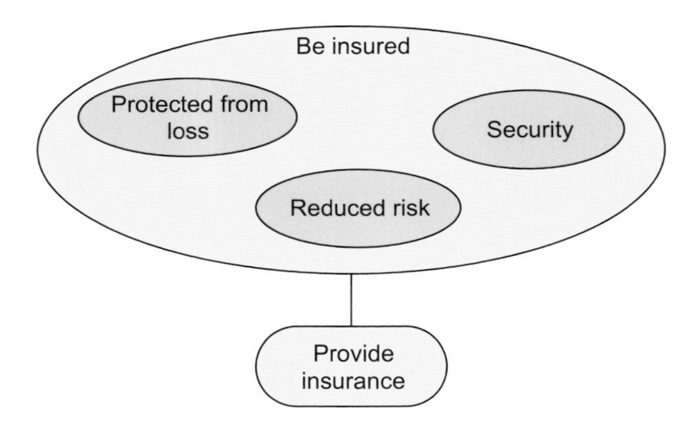

Example 14: Value

3.4.4 Product

A product is defined as a coherent collection of services, accompanied by a contract/set of agreements, which is offered as a whole to (internal or external) customers.

This definition describes financial, services-based, or information products that are common in information-intensive organizations, rather than physical products. A financial or information product consists of a collection of services, and a contract that specifies the characteristics, rights, and requirements associated with the product. "Buying" a product gives the customer the right to use the associated services. Generally, the product concept is used to specify a product *type*. The number of product types in an organization is typically relatively stable compared to, for example, the processes that realize or support the products. "Buying" is usually one of the services associated with a product, which results in a new instance of that product (belonging to a specific customer). Similarly, there may be services to modify or destroy a product.

A product may aggregate business services or application services,[2] as well as a contract. A value may be associated with a product. The name of a product is usually the name which is used in the communication with customers, or possibly a more generic noun (e.g., "travel insurance").

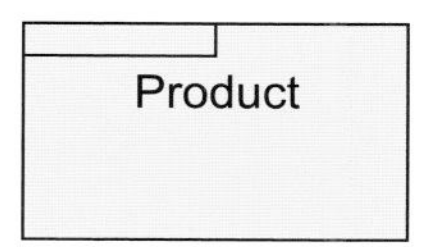

Figure 24: Product Notation

Example

In the model below, a bank offers the product **Telebanking account** to its customers. Opening an account as well as application support (i.e., helpdesk and the like), are modeled as business services realized by the **Customer relations department**. As part of the product, the customer can make use of a banking service which offers application services realized by the **Telebanking application**, such as electronic **Money transfer** and requesting **Account status**.

2 The latter relation is defined in Chapter 6 on cross-layer dependencies.

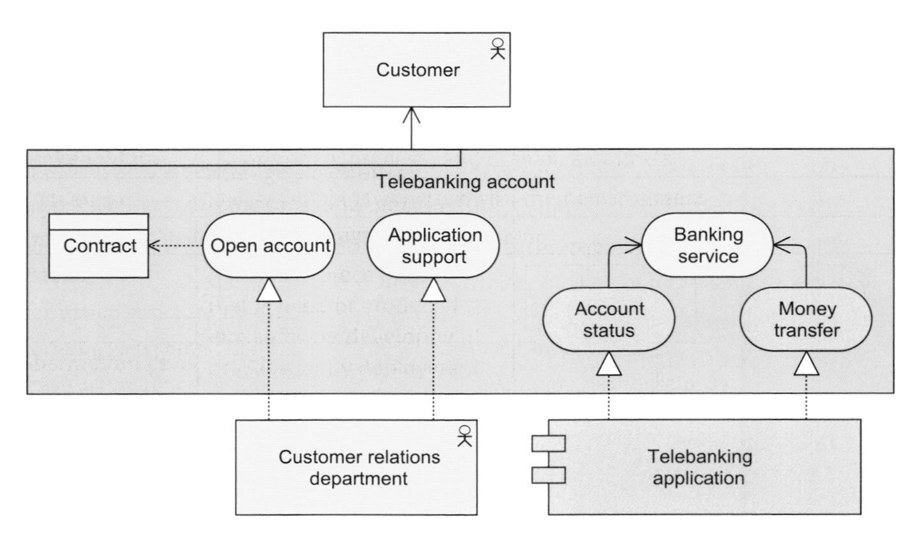

Example 15: Product

3.4.5 Contract

> A contract is defined as a formal or informal specification of an agreement that specifies the rights and obligations associated with a product.

The contract concept may be used to model a contract in the legal sense, but also a more informal agreement associated with a product. It may also be or include a Service Level Agreement (SLA), describing an agreement about the functionality and quality of the services that are part of a product. A contract is a specialization of a business object.

The relationships that apply to a business object also apply to a contract. In addition, a contract may have an aggregation relationship with a product. The name of a contract is preferably a noun.

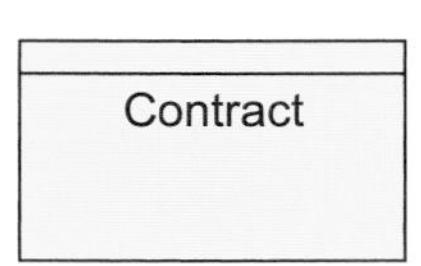

Figure 25: Contract Notation

Example

The model below shows a **Telebanking contract** associated with the product **Telebanking account**. The contract consists of two parts (subcontracts): the **Service Conditions** and a **Service Level Agreement**.

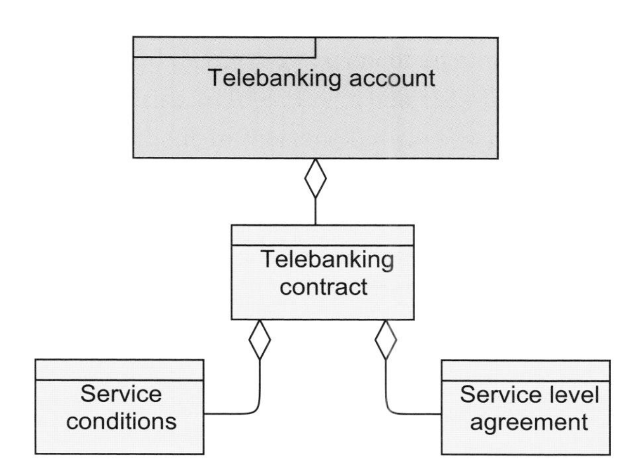

Example 16: Contract

3.5 Summary of Business Layer Concepts

Table 1 gives an overview of the concepts at the business layer, with their definitions.

Table 1: Business Layer Concepts

Concept	Description	Notation
Business actor	An organizational entity that is capable of performing behavior.	Business actor
Business role	The responsibility for performing specific behavior, to which an actor can be assigned.	Business role
Business collaboration	An aggregate of two or more business roles that work together to perform collective behavior.	Business collaboration
Business interface	A point of access where a business service is made available to the environment.	Business interface
Location	A conceptual point or extent in space.	Location
Business object	A passive element that has relevance from a business perspective.	Business object
Business process	A behavior element that groups behavior based on an ordering of activities. It is intended to produce a defined set of products or business services.	Business process
Business function	A behavior element that groups behavior based on a chosen set of criteria (typically required business resources and/or competences).	Business function

Concept	Description	Notation
Business interaction	A behavior element that describes the behavior of a business collaboration.	Business interaction
Business event	Something that happens (internally or externally) and influences behavior.	Business event
Business service	A service that fulfills a business need for a customer (internal or external to the organization).	Business service
Represen-tation	A perceptible form of the information carried by a business object.	Representation
Meaning	The knowledge or expertise present in a business object or its representation, given a particular context.	Meaning
Value	The relative worth, utility, or importance of a business service or product.	Value
Product	A coherent collection of services, accompanied by a contract/set of agreements, which is offered as a whole to (internal or external) customers.	Product
Contract	A formal or informal specification of agreement that specifies the rights and obligations associated with a product.	Contract

Chapter 4

Application Layer

4.1 Application Layer Metamodel

Figure 26 gives an overview of the application layer concepts and their relationships. Many of the concepts have been inspired by the UML 2.0 standard [7], [10], as this is the dominant language and the *de facto* standard for describing software applications. Whenever applicable, we draw inspiration from the analogy with the business and application layer.

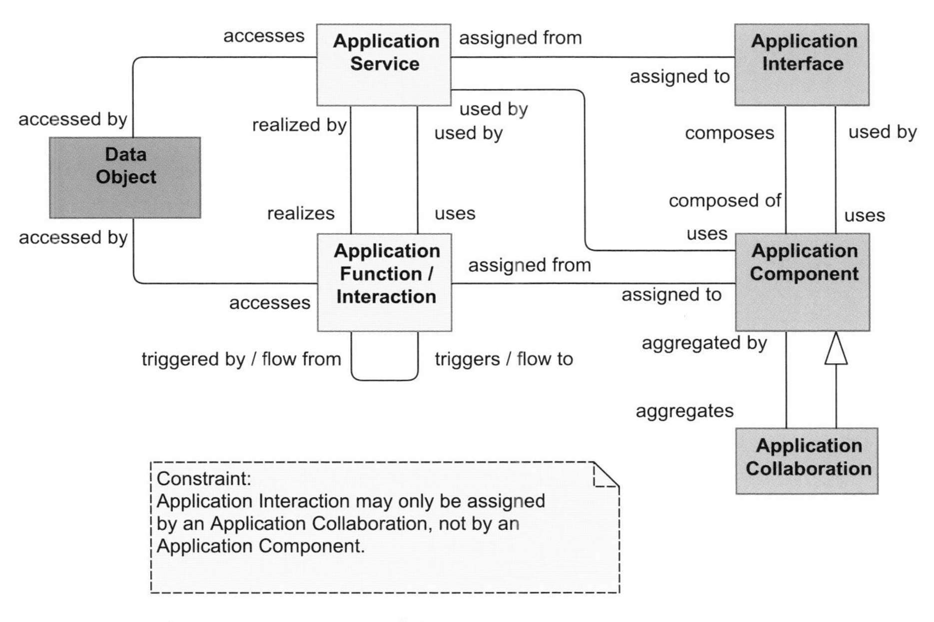

Figure 26: Application Layer Metamodel

Note: This figure does not show all permitted relationships: every concept in the language can have composition, aggregation, and specialization relationships with concepts of the same type; furthermore, there are indirect relationships that can be derived as explained in Section 6.5.

4.2 Structural Concepts

The main structural concept for the application layer is the *application component*. This concept is used to model any structural entity in the application layer: not just (re-usable) software components that can be part of one or more applications, but also complete software applications, sub-applications, or information systems. Although very similar to the UML 2.0 component, our component concept strictly models the structural aspect of an application: its behavior is modeled by an explicit relationship to the behavioral concepts.

Also in application architecture, the inter-relationships of components are an essential ingredient. Therefore, we also introduce the concept of *application collaboration* here, defined as a collective of application components which perform application interactions. The concept is very similar to the collaboration as defined in the UML 2.0 standard [7], [10].

In the purely structural sense, an *application interface* is the (logical) channel through which the services of a component can be accessed. In a broader sense (as used in, among others, the UML 2.0 definition), an application interface defines some elementary behavioral characteristics: it defines the set of operations and events that are provided by the component, or those that are required from the environment. Thus, it is used to describe the functionality of a component. A distinction may be made between a *provided interface* and a *required interface*. The application interface concept can be used to model both *application-to-application* interfaces, which offer internal application services, and *application-to business* interfaces (and/or *user interfaces*), which offer external application services.

Also at the application layer, we distinguish the passive counterpart of the component, which we call a *data object*. This concept is used in the same way as data objects (or object types) in well-known data modeling approaches, most notably the "class" concept in UML class diagrams. A data object can be seen as a representation of a business object, as a counterpart of the representation concept in the business layer.

4.2.1 Application Component

An application component is defined as a modular, deployable, and replaceable part of a software system that encapsulates its behavior and data and exposes these through a set of interfaces.

An application component is a self-contained unit of functionality. As such, it is independently deployable, re-usable, and replaceable. An application component performs one or more application functions. It encapsulates its contents: its functionality is only accessible through a set of application interfaces. Co-operating application components are connected via application collaborations.

An application component may be assigned to one or more application functions, business processes, or business functions. An application component has one or more application interfaces, which expose its functionality. Application interfaces of other application components may be used by an application component. The name of an application component should preferably be a noun.

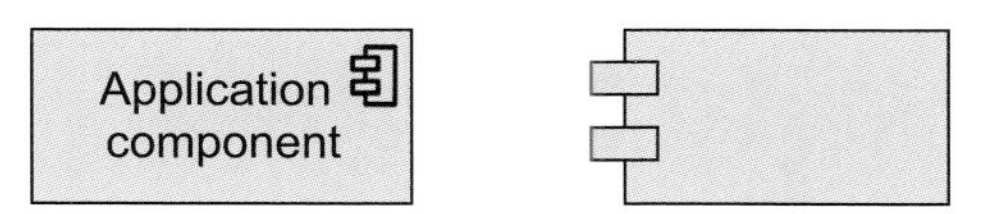

Figure 27: Application Component Notation

Example

In the model below, a financial application is depicted as an application component consisting of two subcomponents for accounting and billing, each of which offers an application service to the environment. These services are accessible through a shared accounting & billing application interface, which is part of the financial application.

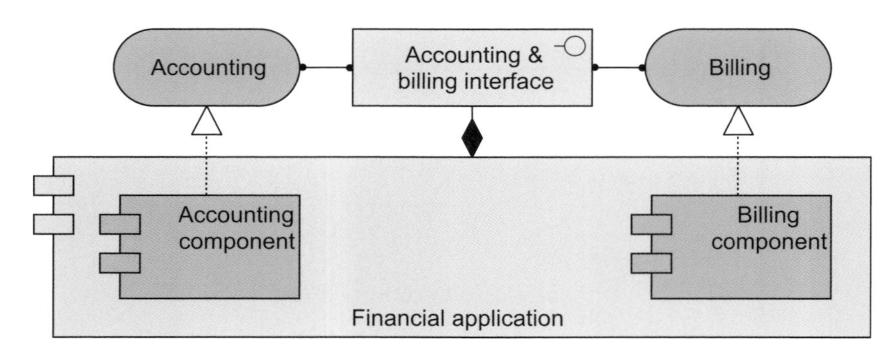

Example 17: Application Component

4.2.2 Application Collaboration

> An application collaboration is defined as an aggregate of two or more application components that work together to perform collective behavior.

An application collaboration specifies which components co-operate to perform some task. The collaborative behavior, including, for example, the communication pattern of these components, is modeled by an application interaction. An application collaboration typically models a logical or temporary collaboration of application components, and does not exist as a separate entity in the enterprise.

An application collaboration is a specialization of a component, and aggregates two or more (co-operating) application components. An application collaboration is an active structure element that may be assigned to one or more application interactions or business interactions, which model the associated behavior. An application interface may be used by an application collaboration, and an application collaboration may be composed of application interfaces. The name of an application collaboration should preferably be a noun.

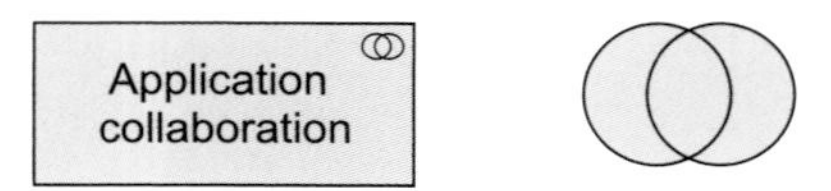

Figure 28: Application Collaboration Notation

> **Example**
> In the model below, two components collaborate in transaction administration: an **Accounting component** and a **Billing component.** This collaboration performs the application interaction **Administrate transactions.**

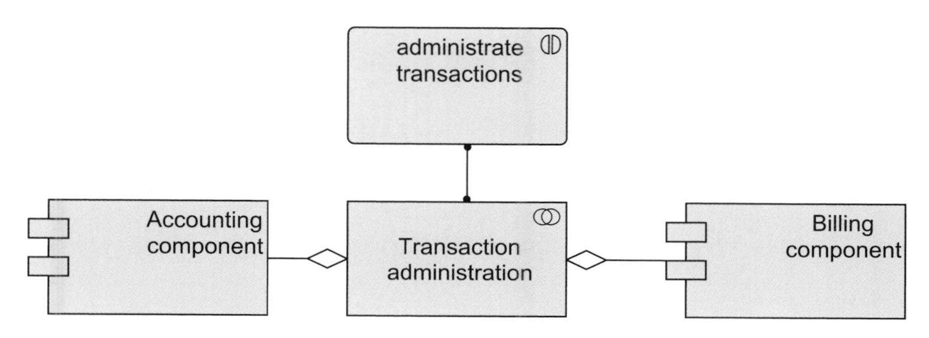

Example 18: Application Collaboration

4.2.3 Application Interface

> An application interface is defined as a point of access where an application service is made available to a user or another application component.

An application interface specifies how the functionality of a component can be accessed by other components (provided interface), or which functionality the component requires from its environment (required interface). An application interface exposes an application service to the environment. The same application service may be exposed through different interfaces.

In a sense, an application interface specifies a kind of contract that a component realizing this interface must fulfill. This may include parameters, protocols used, pre- and post-conditions, and data formats.

An application interface may be part of an application component through composition (not shown in the standard notation), which means that these interfaces are provided or required by that component, and can be used by other application components. An application interface can be assigned to application services or business services, which means that the interface exposes these services to the environment. The name of an application interface should preferably be a noun.

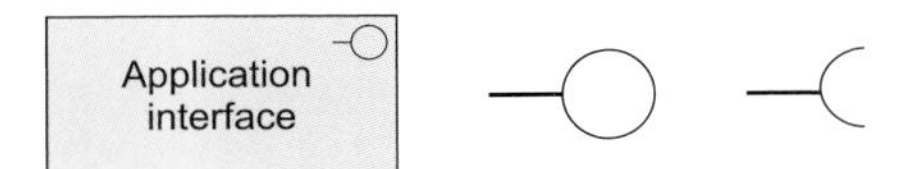

Figure 29: Application Interface Notation

> **Example**
> In the model below, an **Accounting component** is shown that provides an application interface for **Transaction data exchange,** and a **Billing component** that requires such an interface.

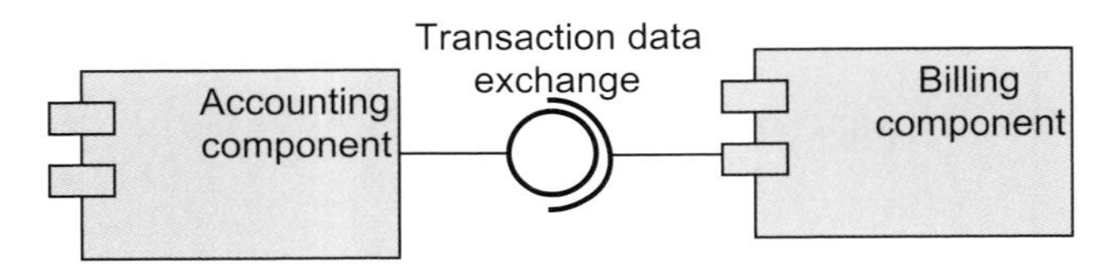

Example 19: Application Interface

4.2.4 Data Object

> A data object is defined as a passive element suitable for automated processing.

An application function operates on a data object. A data object may be communicated via interactions and used or produced by application services. It should be a self-contained piece of information with a clear meaning to the business, not just to the application level. Typical examples of data objects are a customer record, a client database, or an insurance claim.

A data object can be accessed by an application function, application interaction, or application service. A data object may realize a business object, and may be realized by an artifact. A data object may have association, specialization, aggregation, or composition relationships with other data objects. The name of a data object should preferably be a noun.

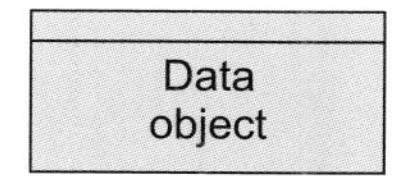

Figure 30: Data Object Notation

Example

In the model below, two application functions co-operate via an application service, in which a data object holding **Transaction data** is exchanged.

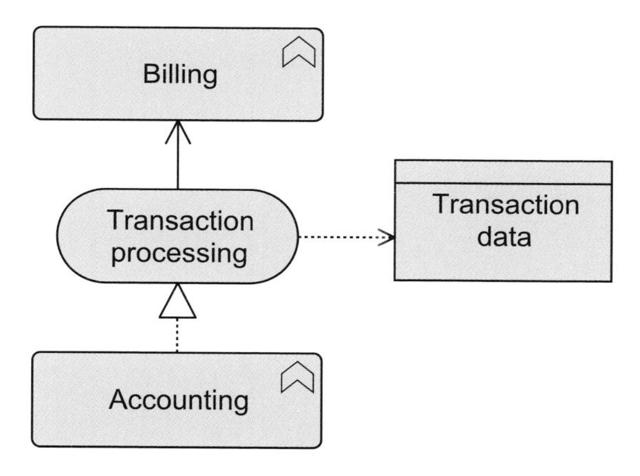

Example 20: Data Object

4.3 Behavioral Concepts

Behavior at the application layer can be described in a way that is very similar to business layer behavior. Also here, we make a distinction between the external behavior of application components in terms of *application services*, and the internal behavior of these components; i.e., *application functions* that realize these services.

An *application service* is an externally visible unit of functionality, provided by one or more components, exposed through well-defined interfaces, and meaningful to the environment. The service concept provides a way to explicitly describe the functionality that components share with each other and the functionality that they make available to the environment. The concept fits well within the current developments in the area of web services. The functionality that an interactive computer program provides through a user interface is also modeled using an application service, exposed by an application-to-business interface representing the user interface. Internal application services are exposed through an application-to-application interface.

An *application function* describes the internal behavior of a component needed to realize one or more application services. In analogy with the business layer, a separate "application flow" concept is conceivable as the counterpart of a business process. Note that the internal behavior of a component should in most cases not be modeled in too much detail in an architectural description, because for the description of this behavior we may soon be confronted with detailed design issues.

An *application interaction* is the behavior of a collaboration of two or more application components. An application interaction is external behavior from the perspective of each of the participating components, but the behavior is internal to the collaboration as a whole.

4.3.1 Application Function

> An application function is defined as a behavior element that groups automated behavior that can be performed by an application component.

An application function describes the internal behavior of an application component. If this behavior is exposed externally, this is done through one or more services. An application function abstracts from the way it is implemented. Only the necessary behavior is specified.

An application function may realize one or more application services. Application services of other application functions and infrastructure services may be used by an application function. An application function may access data objects. An application component may be assigned to an application function (which means that the application component performs the application function). The name of an application function should preferably be a verb ending with "-ing"; e.g., "accounting".

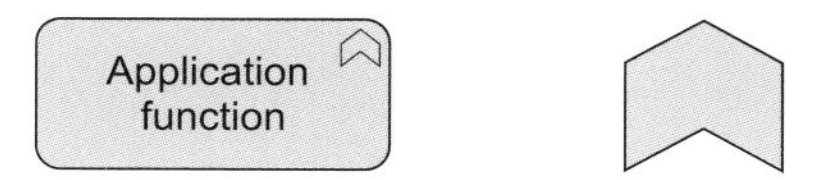

Figure 31: Application Function Notation

Example

In the model below, the internal behavior of the **Financial application** component is modeled as an application function consisting of two sub-functions. These application functions realize the application services that are made available to the users of the application.

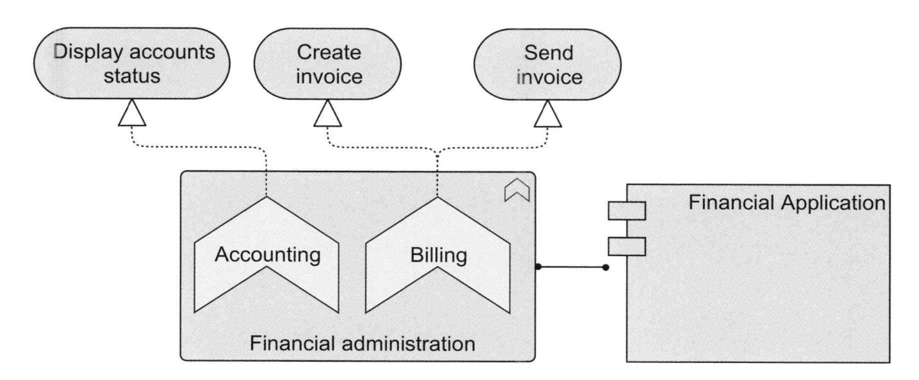

Example 21: Application Function

4.3.2 Application Interaction

An application interaction is defined as a behavior element that describes the behavior of an application collaboration.

An application interaction describes the collective behavior that is performed by the components that participate in an application collaboration. This may, for example, include the communication pattern between these components.

An application interaction can also specify the externally visible behavior needed to realize an application service. The details of the interaction between the application components involved in an application interaction can be expressed during the detailed application design using, e.g., a UML interaction diagram.

An application collaboration may be assigned to an application interaction. An application interaction may realize an application service. Application services and infrastructure services may be used by an application interaction. An application interaction may access data objects. The name of an application interaction should preferably be a verb.

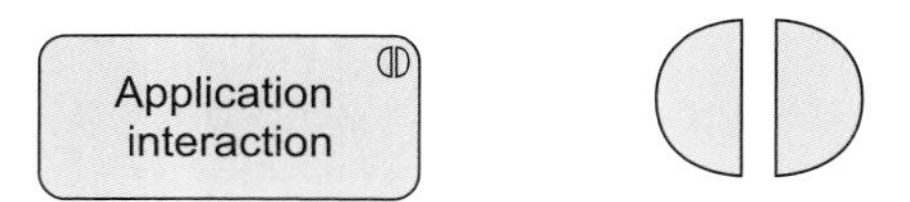

Figure 32: Application Interaction Notation

Example

In the model below, an **Accounting component** and a **Billing component** of a financial system co-operate to compose an *administrate transactions* interaction. This is modeled as an application interaction assigned to the collaboration between the two components.

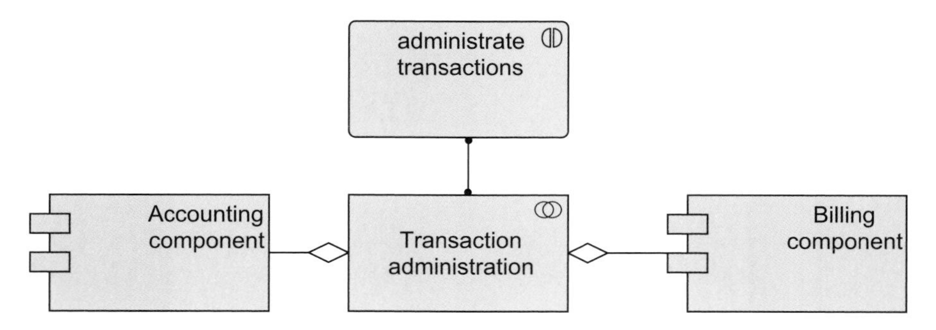

Example 22: Application Interaction

4.3.3 Application Service

> An application service is defined as a service that exposes automated behavior.

An application service exposes the functionality of components to their environment. This functionality is accessed through one or more application interfaces. An application service is realized by one or more application functions that are performed by the component. It may require, use, and produce data objects.

An application service should be meaningful from the point of view of the environment; it should provide a unit of functionality that is, in itself, useful to its users. It has a purpose, which states this utility to the environment. This means, for example, that if this environment includes business processes, application services should have business relevance.

A purpose may be associated with an application service. An application service may be used by business processes, business functions, business interactions, or application functions. An application function may realize an application service. An application interface may be assigned to an application service. An application service may access data objects. The name of an application service should preferably be a verb ending with "-ing"; e.g., "transaction processing". Also, a name explicitly containing the word "service" may be used.

Application
service

Figure 33: Application Service Notation

Example

In the model below, a **Transaction processing** (application-to-application) service is realized by the **Accounting** application function, and is accessible by other components through a **Transaction processing** application programming interface (API). This service is used by the **Billing** application function performed by the **Billing** component.

The **Billing** application function offers an (application-to-business) function **Bill creation**, which can be used to support business processes, and is accessible to business roles through a **Billing screen** as an application-to-business interface.

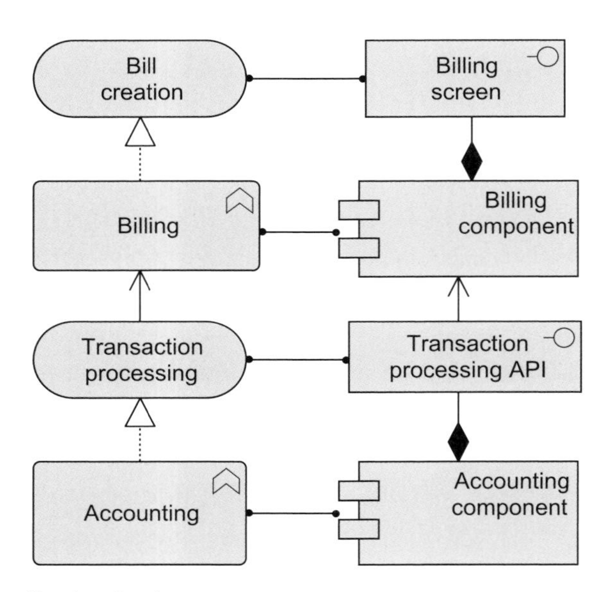

Example 23: Application Service

4.4 Summary of Application Layer Components

Table 2 gives an overview of the concepts at the application layer, with their definitions.

Table 2: Application Layer Concepts

Concept	Definition	Notation
Application component	A modular, deployable, and replaceable part of a software system that encapsulates its behavior and data and exposes these through a set of interfaces.	Application component
Application collaboration	An aggregate of two or more application components that work together to perform collective behavior.	Application collaboration
Application interface	A point of access where an application service is made available to a user or another application component.	Application interface
Data object	A passive element suitable for automated processing.	Data object
Application function	A behavior element that groups automated behavior that can be performed by an application component.	Application function
Application interaction	A behavior element that describes the behavior of an application collaboration.	Application interaction
Application service	A service that exposes automated behavior.	Application service

Chapter 5

Technology Layer

5.1 Technology Layer Metamodel

Figure 34 gives an overview of the technology layer concepts and their relationships. Many of the concepts have been inspired by the UML 2.0 standard [7], [10], as this is the dominant language and the *de facto* standard for describing software applications and infrastructure. Whenever applicable, we draw inspiration from the analogy with the business and application layers.

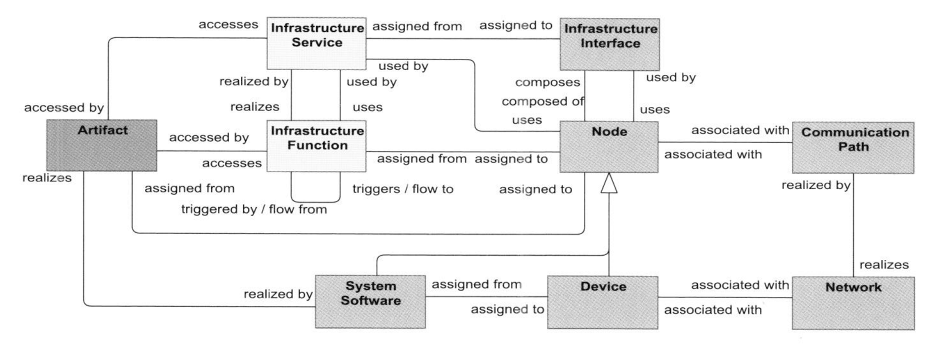

Figure 34: Technology Layer Metamodel

Note: This figure does not show all permitted relationships: every concept in the language can have composition, aggregation, and specialization relationships with concepts of the same type; furthermore, there are indirect relationships that can be derived as explained in Section 6.5.

5.2 Structural Concepts

The main structural concept for the technology layer is the *node*. This concept is used to model structural entities in this layer. It is identical to the node concept of UML 2.0. It strictly models the structural aspect of a system: its behavior is modeled by an explicit relationship to the behavioral concepts.

An *infrastructure interface* is the (logical) location where the infrastructure services offered by a node can be accessed by other nodes or by application components from the application layer.

Nodes come in two flavors: *device* and *system software*, both taken from UML 2.0. A *device* models a physical computational resource, upon which artifacts may be deployed for execution. *System software* is an infrastructural software component running on a device. Typically, a node consists of a number of sub-nodes; for example, a device such as a server and system software to model the operating system.

The inter-relationships of components in the technology layer are mainly formed by the communication infrastructure. The *communication path* models the relation between two or more nodes, through which these nodes can exchange information. The physical realization of a communication path is a modeled with a *network*; i.e., a physical communication medium between two or more devices (or other networks).

5.2.1 Node

> A node is defined as a computational resource upon which artifacts may be stored or deployed for execution.

Nodes are active processing elements that execute and process artifacts, which are the representation of components and data objects. Nodes are, for example, used to model application servers, database servers, or client workstations. A node is often a combination of a hardware device and system software, thus providing a complete execution environment. These sub-nodes that represent the hardware devices and system software may be modeled explicitly or left implicit.

Nodes can be interconnected by communication paths. Artifacts can be assigned to (i.e., deployed on) nodes.

The name of a node should preferably be a noun. A node can consist of sub-nodes.

Artifacts deployed on a node may either be drawn inside the node or connected to it with an assignment relationship.

Figure 35: Node Notation

Example

In the model below, we see an **Application Server** node, which consists of a **Blade** device and **Java EE-based application server** system software.

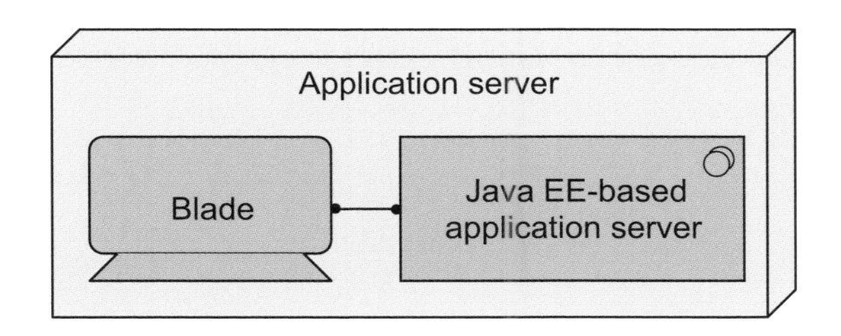

Example 24: Node

5.2.2 Device

A device is defined as a hardware resource upon which artifacts may be stored or deployed for execution.

A device is a specialization of a node that represents a physical resource with processing capability. It is typically used to model hardware systems such as mainframes, PCs, or routers. Usually, they are part of a node together with system software. Devices may be composite; i.e., consist of sub-devices.

Devices can be interconnected by networks. Artifacts can be assigned to (i.e., deployed on) devices. System software can be assigned to a device. A node can contain one or more devices.

The name of a device should preferably be a noun referring to the type of hardware; e.g., "IBM System z mainframe".

A device can consist of sub-devices.

Different icons may be used to distinguish between different types of devices; e.g. mainframes and PCs.

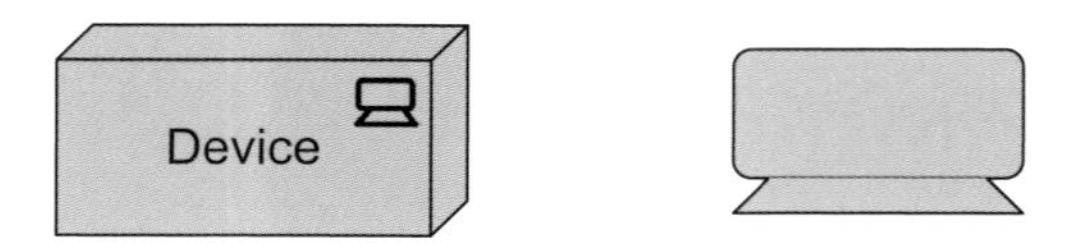

Figure 36: Device Notation

> **Example**
> The model below shows an example of a number of servers, modeled as devices, interconnected through a local area network (LAN).

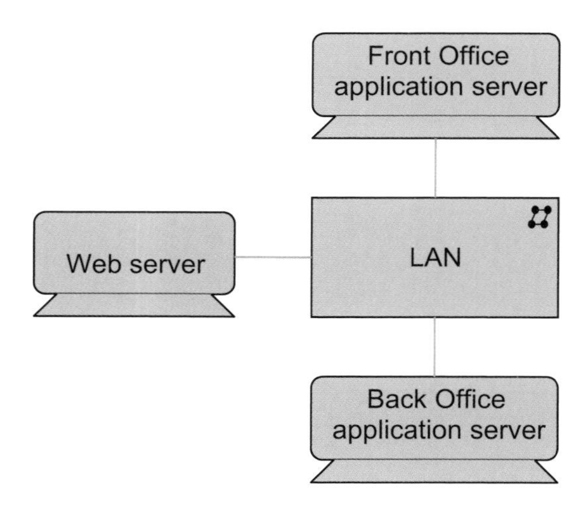

Example 25: Device

5.2.3 System Software

System software represents a software environment for specific types of components and objects that are deployed on it in the form of artifacts.

System software is a specialization of a node that is used to model the software environment in which artifacts run. This can be, for example, an operating system, a JEE application server, a database system, a workflow engine, or COTS software such as ERP or CRM packages. Also, system software can be used to represent, for example, communication middleware. Usually, system software is combined with a device representing the hardware environment to form a general node.

System software can be assigned to a device. Artifacts can be assigned to (i.e., deployed on) system software. A node can contain system software.

The name of system software should preferably be a noun referring to the type of execution environment; e.g., "JEE server". System software may contain other system software; e.g., an operating system containing a database.

Figure 37: System Software Notation

Example
In the model below, we see a **mainframe** device that deploys two system software environments: a customer transaction server and a database management system (DBMS).

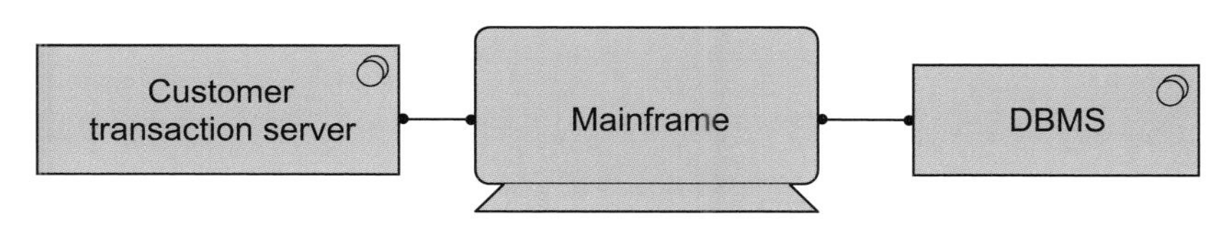

Example 26: System Software

5.2.4 Infrastructure Interface

An infrastructure interface is defined as a point of access where infrastructure services offered by a node can be accessed by other nodes and application components.

An infrastructure interface specifies how the infrastructure services of a node can be accessed by other nodes (provided interface), or which functionality the node requires from its environment (required interface). An infrastructure interface exposes an infrastructure service to the environment. The same service may be exposed through different interfaces.

In a sense, an infrastructure interface specifies a kind of contract that a component realizing this interface must fulfill. This may include, for example, parameters, protocols used, pre- and post-conditions, and data formats.

An infrastructure interface may be part of a node through composition (not shown in the standard notation), which means that these interfaces are provided or required by that node, and can be used by other nodes. An infrastructure service can be assigned to an infrastructure interface, which exposes the service to the environment.

The name of an infrastructure interface should preferably be a noun.

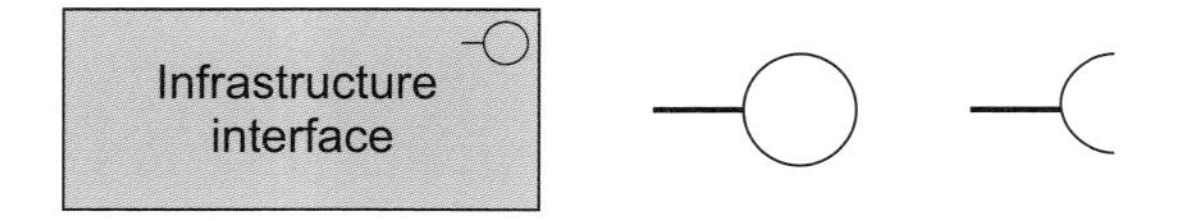

Figure 38: Infrastructure Interface Notations

Example

In the model below, we see a client infrastructure interface exposed, which is part of the client-server system software.

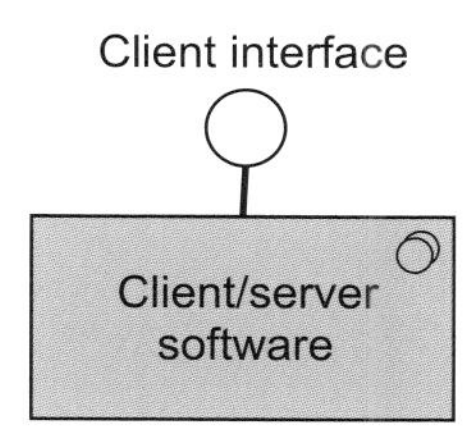

Example 27: Infrastructure Interface

5.2.5 Network

A network is defined as a communication medium between two or more devices.

A network represents the physical communication infrastructure. This may comprise one or more fixed or wireless network links. The most basic network is a single link between two devices. A network has properties such as bandwidth and latency. It embodies the physical realization of the logical communication paths between nodes.

A network connects two or more devices. A network realizes one or more communication paths.

A network can consist of sub-networks.

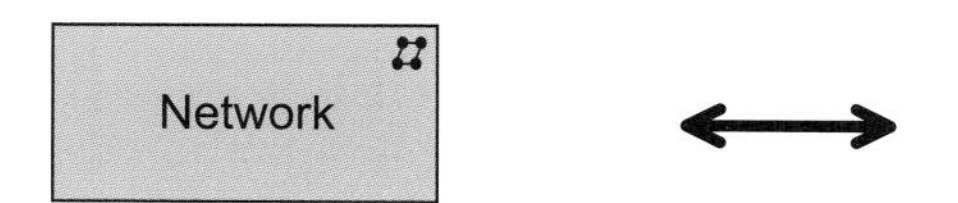

Figure 39: Network Notation, as Connection and as Box

Example
In the model below, a 100 Mb/s LAN network connects a mainframe and PC device.

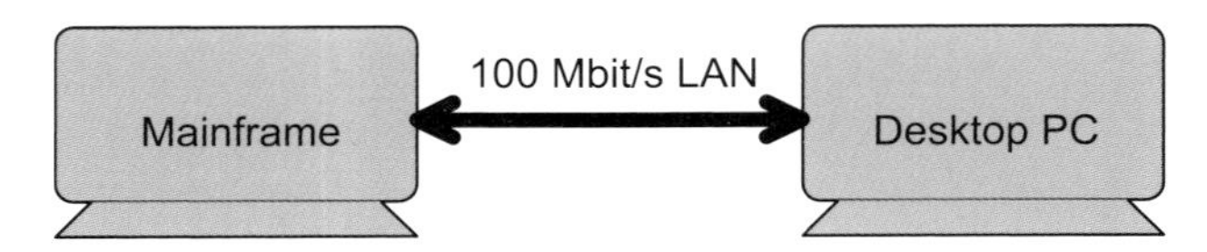

Example 28: Network

5.2.6 Communication Path

> A communication path is defined as a link between two or more nodes, through which these nodes can exchange data.

A communication path is used to model the logical communication relations between nodes. It is realized by one or more networks, which represent the physical communication links. The communication properties (e.g., bandwidth, latency) of a communication path are usually aggregated from these underlying networks.

A communication path connects two or more nodes. A communication path is realized by one or more networks. A communication path is atomic.

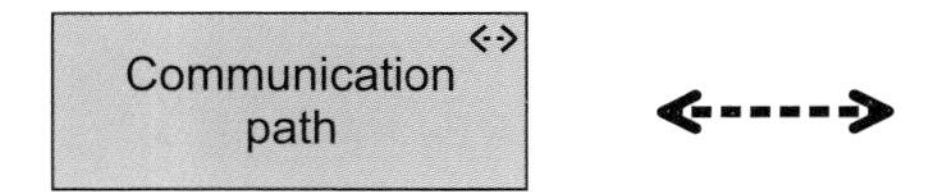

Figure 40: Communication Path Notation, as Connection and as Box

> **Example**
> In the model below, we see a communication path "message queuing" between an **Application Server** and a **Client**.

Example 29: Communication Path

5.3 Behavioral Concepts

Behavior elements in the technology layer are similar to the behavior elements in the other two layers. Also here, we make a distinction between the external behavior of nodes in terms of *infrastructure services*, and the internal behavior of these nodes; i.e., *infrastructure functions* that realize these services.

5.3.1 Infrastructure Function

> An infrastructure function is defined as a behavior element that groups infrastructural behavior that can be performed by a node.

An infrastructure function describes the internal behavior of a node; for the user of a node that performs an infrastructure function, this function is invisible. If its behavior is exposed externally, this is done through one or more infrastructure services. An infrastructure function abstracts from the way it is implemented. Only the necessary behavior is specified.

An infrastructure function may realize infrastructure services. Infrastructure services of other infrastructure functions may be used by an infrastructure function. An infrastructure function may access artifacts. A node may be assigned to an infrastructure function (which means that the node performs the infrastructure function). The name of an infrastructure function should preferably be a verb ending with "-ing".

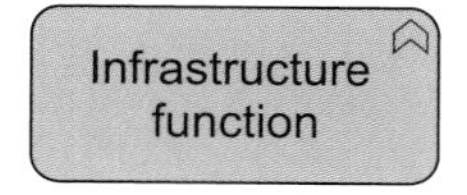

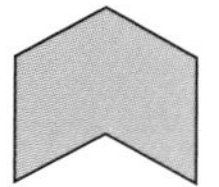

Figure 41: Infrastructure Function Notation

> **Example**
> In the model below, the database management system (DBMS) node performs two infrastructure functions: providing data access (realizing a data access service for application software), and managing data (realizing a data management service for database administration).

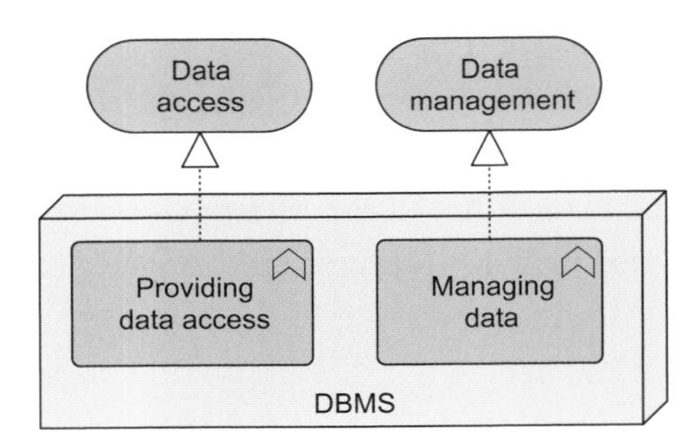

Example 30: Infrastructure Function

5.3.2 Infrastructure Service

> An infrastructure service is defined as an externally visible unit of functionality, provided by one or more nodes, exposed through well-defined interfaces, and meaningful to the environment.

An infrastructure service exposes the functionality of a node to its environment. This functionality is accessed through one or more infrastructure interfaces. It may require, use, and produce artifacts.

An infrastructure service should be meaningful from the point of view of the environment; it should provide a unit of functionality that is, in itself, useful to its users, such as application components and nodes.

Typical infrastructure services may, for example, include messaging, storage, naming, and directory services. It may access artifacts; e.g., a file containing a message.

An infrastructure service may be used by application components or nodes. An infrastructure service is realized by a node. An infrastructure service is exposed by a node by assigning it to its infrastructure interfaces. An infrastructure service may access artifacts.

The name of an infrastructure service should preferably be a verb ending with "-ing"; e.g., "messaging". Also, a name explicitly containing the word "service" may be used.

An infrastructure service may consist of sub-services.

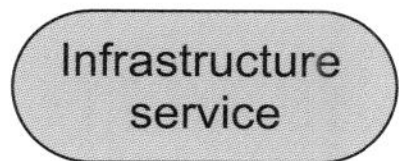

Figure 42: Infrastructure Interface Notation

Example
In the model below, we see a **Messaging service** realized by **Message-Oriented Middleware (MOM)** system software.

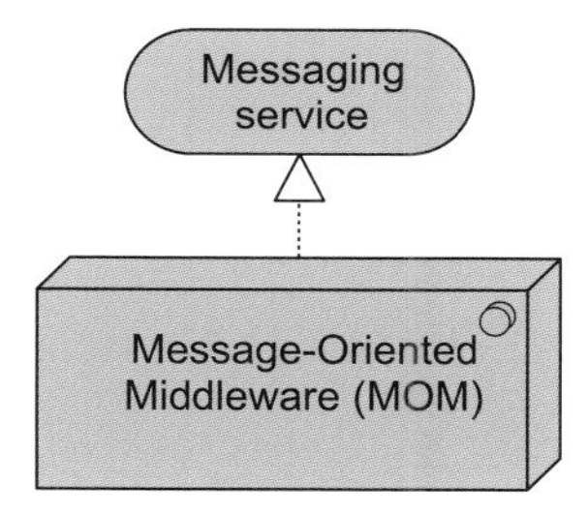

Example 31: Infrastructure Service

5.4 Informational Concepts

An *artifact* is a physical piece of information that is used or produced in a software development process, or by deployment and operation of a system. It is the representation, in the form of, for example, a file, of a data object, or an application component, and can be deployed on a node. The artifact concept has been taken from UML 2.0.

5.4.1 Artifact

An artifact is defined as a physical piece of data that is used or produced in a software development process, or by deployment and operation of a system.

An artifact represents a concrete element in the physical world. It is typically used to model (software) products such as source files, executables, scripts, database tables, messages, documents, specifications, and model files. An instance (copy) of an artifact can be deployed on a node.

An application component or system software may be realized by one or more artifacts. A data object may be realized by one or more artifacts. A node may be assigned to an artifact (i.e., the artifact is deployed on the node). Thus, the two typical ways to use the artifact concept are as an *execution component* or as a *data file*. In fact, these could be defined as specializations of the artifact concept.

The name of an artifact should preferably be the name of the file it represents; e.g., "order.jar". An artifact may consist of sub-artifacts.

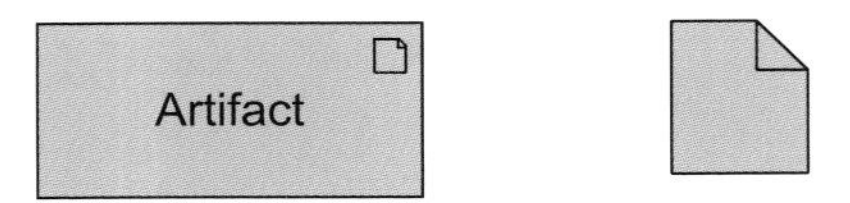

Figure 43: Artifact Notation

> **Example**
> In the example below, we see an artifact **Risk management EJB,** which represents a deployable unit of code, assigned to (deployed on) an application server.

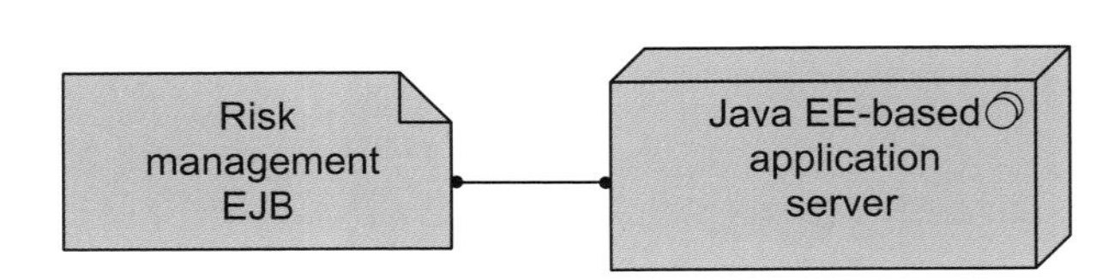

Example 32: Artifact

5.5 Summary of Technology Layer Concepts

Table 3 gives an overview of the concepts at the technology layer, with their definitions.

Table 3: Technology Layer Concepts

Concept	Definition	Notation
Node	A computational resource upon which artifacts may be stored or deployed for execution.	Node
Device	A hardware resource upon which artifacts may be stored or deployed for execution.	Device
Network	A communication medium between two or more devices.	Network
Communication path	A link between two or more nodes, through which these nodes can exchange data.	Communication path
Infrastructure interface	A point of access where infrastructure services offered by a node can be accessed by other nodes and application components.	Infrastructure interface
System software	A software environment for specific types of components and objects that are deployed on it in the form of artifacts.	System software
Infrastructure function	A behavior element that groups infrastructural behavior that can be performed by a node.	Infrastructure function

Concept	Definition	Notation
Infrastructure service	An externally visible unit of functionality, provided by one or more nodes, exposed through well-defined interfaces, and meaningful to the environment.	Infrastructure service
Artifact	A physical piece of data that is used or produced in a software development process, or by deployment and operation of a system.	Artifact

Chapter 6

Cross-Layer Dependencies

In the previous chapters we have presented the concepts to model the business, application, and technology layers of an enterprise. However, a central issue in enterprise architecture is business-IT alignment: how can these layers be matched? In this chapter, we describe the relationships that the ArchiMate language offers to model the link between business, applications, and technology.

6.1 Business-Application Alignment

Figure 44 shows the relationships between the business layer, the application layer, and the technology layer concepts. There are three main types of relationships between these layers:

1. *Used by* relationships, between application service and the different types of business behavior elements, and between application interface and business role. These relationships represent the behavioral and structural aspects of the support of the business by applications.
2. A *realization* relationship from a data object to a business object, to indicate that the data object is a digital representation of the corresponding business object.
3. *Assignment* relationships, between application component and business process, function, or interaction, and between application interface and business service, to indicate that, for example, business processes or business services are completely automated. The case that a business process, function, or interaction is not completely automated but only supported by an application component is expressed with a "used by" relationship (see, e.g., the example of an Application Usage Viewpoint in Section 7.4.11).

In addition, there may be an aggregation relationship between a product and an application or infrastructure service, to indicate that the application or infrastructure service can be offered directly to a customer as part of the product. Also, a location may be assigned to all active and passive structural elements (and, indirectly, behavior elements) in the application and technology layers.

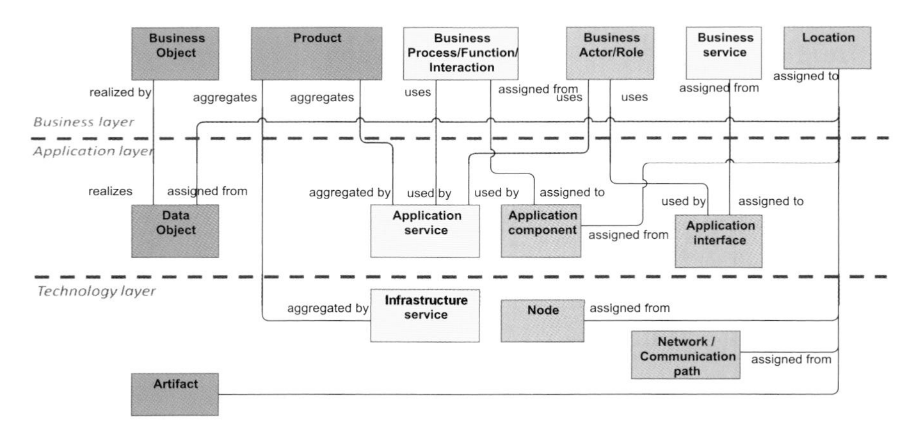

Figure 44: Relationships between Business Layer and Lower Layer Concepts

Note: This figure does not show all permitted relationships: there are indirect relationships that can be derived as explained in Section 6.5.

6.2 Application-Technology Alignment

Figure 45 shows the relationships between application layer and technology layer concepts. There are two types of relationships between these layers:

1. *Used by* relationships, between infrastructure service and the different types of application behavior elements, and between infrastructure interface and application component. These relationships represent the behavioral and structural aspects of the use of technical infrastructure by applications.
2. A *realization* relationship from artifact to data object, to indicate that the data object is realized by, for example, a physical data file, and from artifact to application component, to indicate that a physical data file is an executable that realizes an application or part of an application. (Note: In this case, an artifact represents a "physical" component that is deployed on a node; this is modeled with an assignment relationship. A (logical) application component is realized by an artifact and, indirectly, by the node on which the artifact is deployed.)

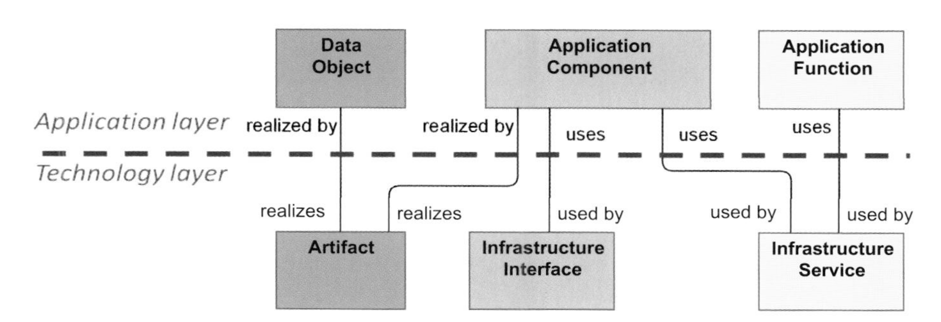

Figure 45: Relationships between Application Layer and Technology Layer Concepts

Note: This figure does not show all permitted relationships: there are indirect relationships that can be derived as explained in Section 6.5.

Due to the derived relationships that are explained in Section 6.5, it is also possible to draw relationships directly between the business and technology layers. For example, if a business object is realized by a data object, which in turn is realized by an artifact, this artifact indirectly realizes the business object.

Chapter 7

Relationships

The metamodels and examples from the previous chapters show the different types of relationships that the ArchiMate language offers. In this chapter, we provide a more precise description of these relationships.

The relationships can be classified as either:

- *Structural*, which model the structural coherence of concepts of the same or different types
- *Dynamic*, which are used to model (temporal) dependencies between behavioral concepts
- *Other*, which do not fall in one of the two above categories

7.1 Structural Relationships

7.1.1 Composition Relationship

> The composition relationship indicates that an object is composed of one or more other objects.

The composition relationship has been inspired by the composition relationship in UML class diagrams. In contrast to the aggregation relationship, an object can be part of only one composition.

In addition to composition relationships that are explicitly defined in the metamodel figures of the previous sections, composition is always possible between two instances of the same concept.

Figure 46: Composition Notation

Alternatively, a composition relationship can be expressed by nesting the model elements.

> **Example**
> The models below show the two ways to express that the application component **Financial application** is composed of three other application components.

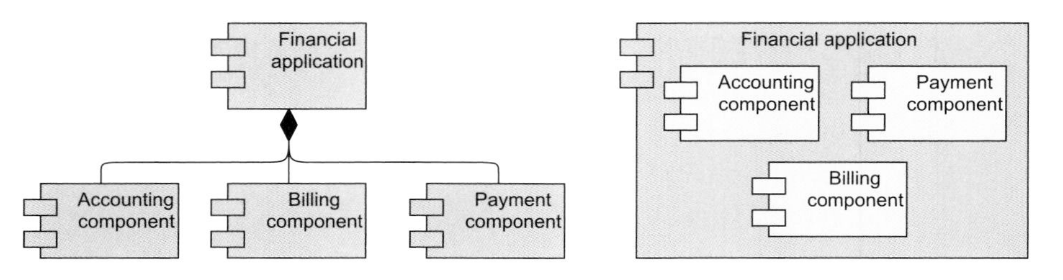

Example 33: Composition

7.1.2 Aggregation Relationship

> The aggregation relationship indicates that a concept groups a number of other concepts.

The aggregation relationship has been inspired on the aggregation relationship in UML class diagrams. In contrast to the composition relationship, an object can be part of more than one aggregation.

In addition to aggregation relationships that are explicitly defined in the metamodel figures of the previous sections, aggregation is always possible between two instances of the same concept.

Figure 47: Aggregation Notation

Alternatively, an aggregation relationship can be expressed by nesting the model elements.

Example

The models below show the two ways to express that the product **Car insurance** aggregates a contract (**Policy**) and two business services.

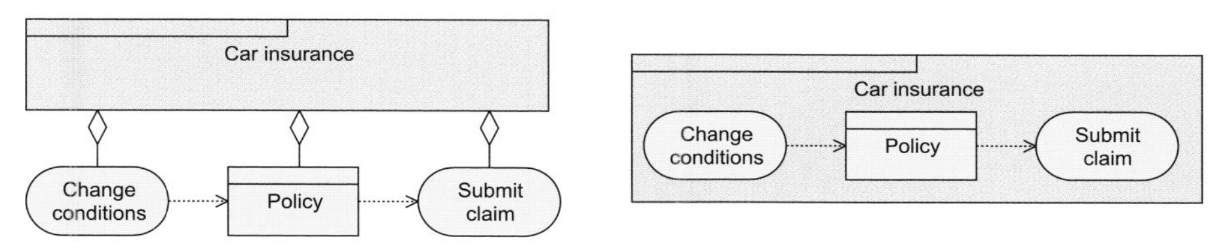

Example 34: Aggregation

7.1.3 Assignment Relationship

The assignment relationship links active elements (e.g., business roles or application components) with units of behavior that are performed by them, or business actors with business roles that are fulfilled by them.

The assignment relationship can relate a business role with a business process or function, an application component with an application function, a business collaboration with a business interaction, an application collaboration with an application interaction, a business interface with a business service, an application interface with an application service, or a business actor with a business role.

Figure 48: Assignment Notation

Alternatively, an assignment relationship can be expressed by nesting the model elements.

> **Example**
>
> The model in the example below includes the two ways to express the assignment relationship. The **Payment function** (application) is assigned to the **Financial application** (component), and the **Payment service** (application) is assigned to the **Application interface**.

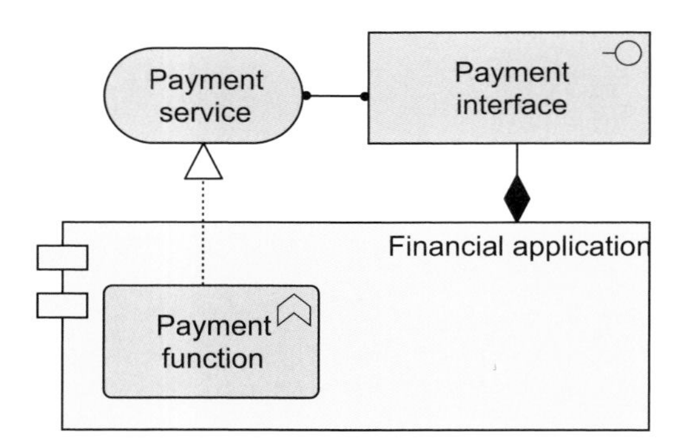

Example 35: Assignment

7.1.4 Realization Relationship

> The realization relationship links a logical entity with a more concrete entity that realizes it.

The realization relationship indicates how logical entities ("what"), such as services, are realized by means of more concrete entities ("how"). The realization relationship is used in an operational sense (e.g., a process/ function realizes a service), but also in a design/implementation context (e.g., a data object may realize a business object, or an artifact may realize an application component).

Figure 49: Realization Notation

Example

The model below illustrates two ways to use the realization relationship. An application (component) **Financial application** realizes the **Billing service** (application); the **Billing data** object realizes the business object **Invoice.**

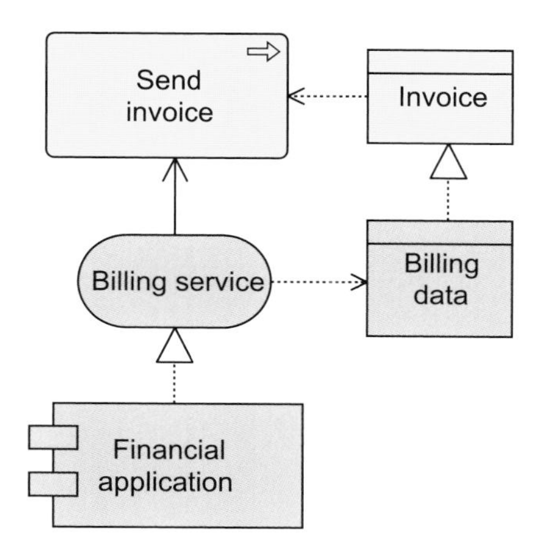

Example 36: Realization

7.1.5 Used By Relationship

The used by relationship models the use of services by processes, functions, or interactions and the access to interfaces by roles, components, or collaborations.

The used by relationship describes the services that a role or component offers that are used by entities in the environment. The used by relationship is applied for both the behavior aspect and the structure aspect. (Note that, although the notation of the "used by" relationship resembles the notation of the dependency relationship in UML, the relationship has a distinct meaning in ArchiMate.)

Figure 50: Used By Notation

Example

The model below illustrates the used by relationship: an application interface (in this case, the user interface of the application) is used by the **Front office employee**, while the **Update customer info** service is used in the **Process change of address** business process.

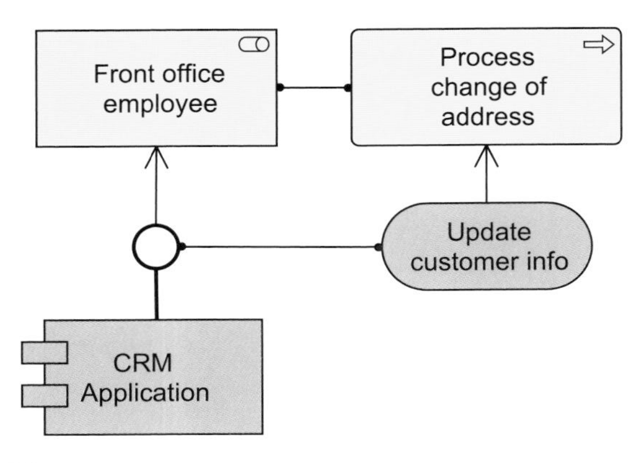

Example 37: Used By

7.1.6 Access Relationship

The access relationship models the access of behavioral concepts to business or data objects.

The access relationship indicates that a process, function, interaction, service, or event "does something" with a (business or data) object; e.g., create a new object, read data from the object, write or modify the object data, or delete the object. The relationship can also be used to indicate that the object is just associated with the behavior; e.g., it models the information that comes with an event, or the information that is made available as part of a service. The arrow head, if present, indicates the direction of the flow of information. (The access relationship should not be confused with the UML dependency relationship, which uses a similar notation.)

Figure 51: Access Notation

> **Example**
> The model below illustrates the access relationship: the **Create invoice** sub-process writes/creates the **Invoice** business object; the **Send invoice** sub-process reads the **Invoice** business object.

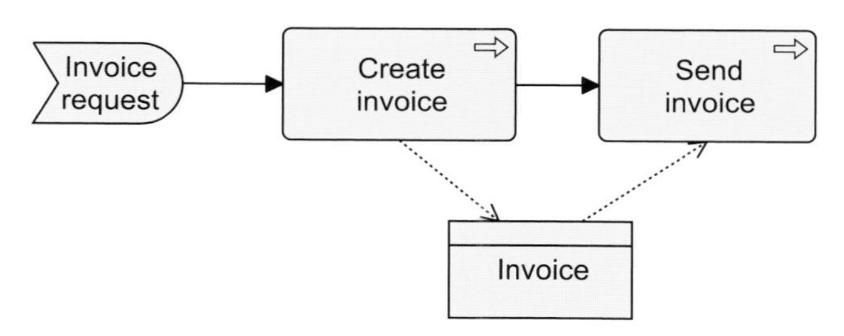

Example 38: Access

7.1.7 Association Relationship

> An association models a relationship between objects that is not covered by another, more specific relationship.

Association is mainly used, as in UML, to model relationships between business objects or data objects that are not modeled by the standard relationships aggregation, composition, or specialization. In addition to this, the association relationship is used to link the informational concepts with the other concepts: a business object with a representation, a representation with a meaning, and a business service with a purpose.

Figure 52: Association Notation

> **Example**
> The model illustrates a number of uses of the association relationship.

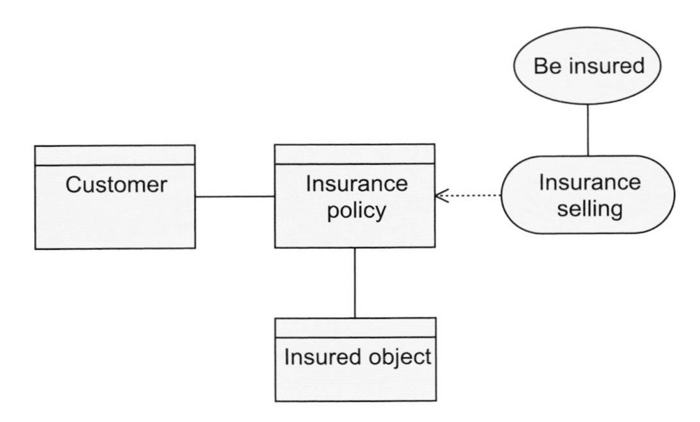

Example 39: Association

7.2 Dynamic Relationships

7.2.1 Triggering Relationship

> The triggering relationship describes the temporal or causal relationships between processes, functions, interactions, and events.

The triggering relationship is used to model the causal relationships between behavior concepts in a process. No distinction is made between an active triggering relationship and a passive causal relationship.

Figure 53: Triggering Notation

> **Example**
> The model below illustrates that triggering relationships are mostly used to model causal dependencies between (sub-)processes and/or events.

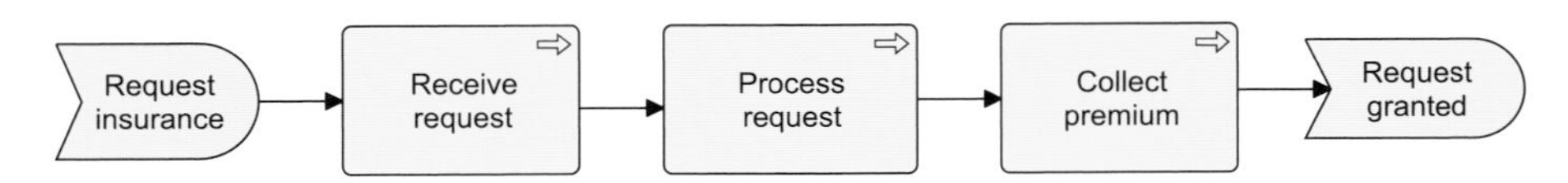

Example 40: Triggering

7.2.2 Flow Relationship

> The flow relationship describes the exchange or transfer of, for example, information or value between processes, function, interactions, and events.

The flow relationship is used to model the flow of, for example, information between behavior concepts in a process. A flow relationship does not imply a causal or temporal relationship.

Figure 54: Flow Notation

> **Example**
> The model below shows a **Claim assessment** business function, which forwards decisions about the claims to the **Claim settlement** business function. In order to determine the order in which the claims should be assessed, **Claim assessment** makes use of schedule information received from the **Scheduling** business function.

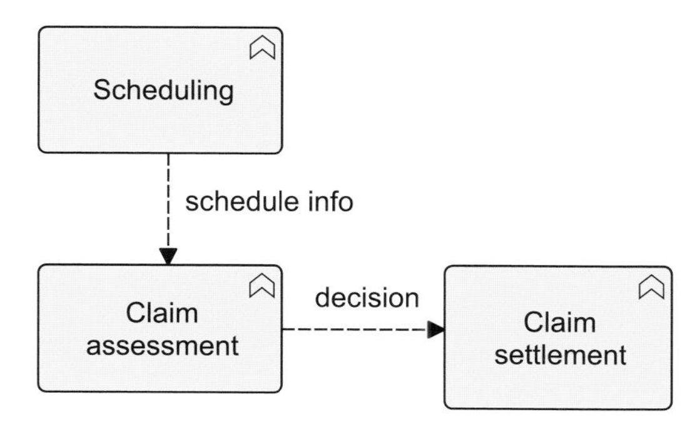

Example 41: Flow

7.3 Other Relationships

7.3.1 Grouping

The grouping relationship indicates that objects belong together based on some common characteristic.

Similar to the UML package, the grouping relationship is used to group an arbitrary group of model objects, which can be of the same type or of different types. In contrast to the aggregation or composition relationships, there is no "overall" object of which the grouped objects form a part.

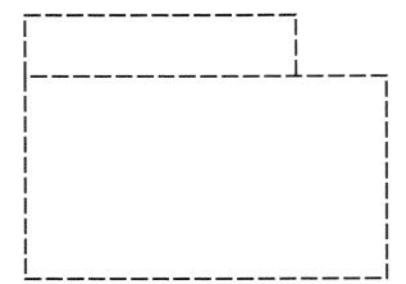

Figure 55: Grouping Notation

Unlike the other language concepts, grouping has no formal semantics. It is only used to show graphically that model elements have something in common. Model elements may belong to multiple (overlapping) groups.

Example
In the model below, the grouping relationship is used to group business objects that belong to the same information domain, in this case **Financial administration**.

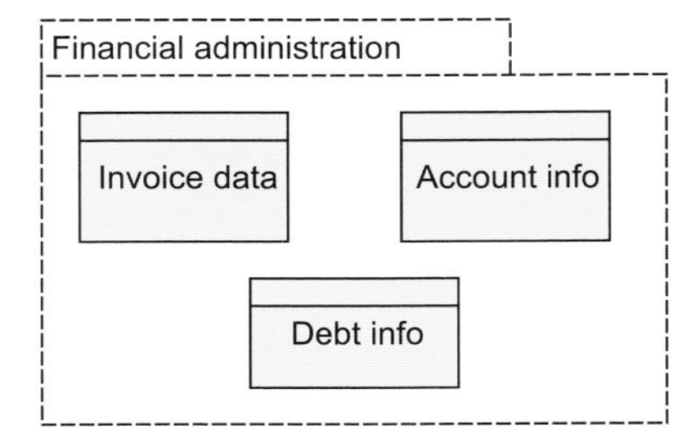

Example 42: Grouping

7.3.2 Junction

A junction is used to connect dynamic relationships of the same type.

A junction is used in a number of situations to connect dynamic (triggering or flow) relationships of the same type; e.g., to indicate splits or joins.

Figure 56: Junction Notation

Example

In the model below, a junction is used to denote an *or*-split (choice).

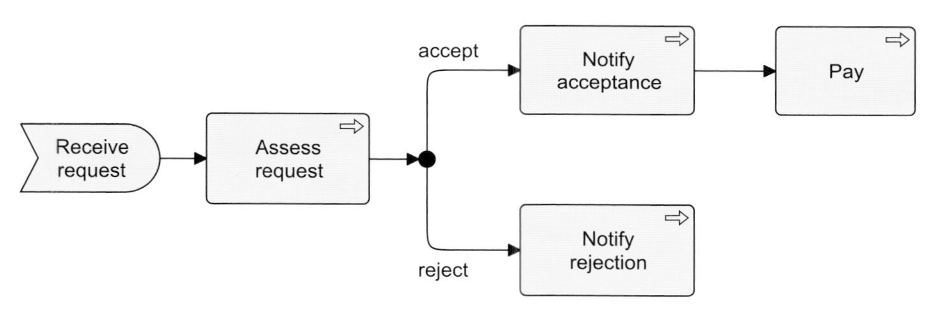

Example 43: Junction

7.3.3 Specialization Relationship

The specialization relationship indicates that an object is a specialization of another object.

The specialization relationship has been inspired by the generalization/specialization relationship in UML class diagrams, but is applicable to specialize a wider range of concepts. The specialization relationship can relate any instance of a concept with another instance of the same concept.

Specialization is always possible between two instances of the same concept.

Figure 57: Specialization Notation

ExampleThe model below illustrates the use of the specialization relationship for a business process. In this case the **Take out travel insurance** and **Take out luggage insurance** processes are a specialization of a more generic insurance take out process.

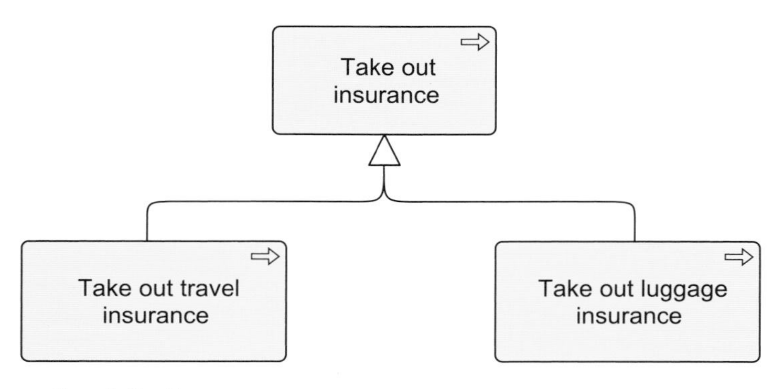

Example 44: Specialization

7.4 Summary of Relationships

Table 4 gives an overview of the ArchiMate relationships with their definitions.

Table 4: Relationships

Structural Relationships		**Notation**
Association	Association models a relationship between objects that is not covered by another, more specific relationship.	————
Access	The access relationship models the access of behavioral concepts to business or data objects.	··········> ············

Structural Relationships		Notation
Used by	The used by relationship models the use of services by processes, functions, or interactions and the access to interfaces by roles, components, or collaborations.	
Realization	The realization relationship links a logical entity with a more concrete entity that realizes it.	
Assignment	The assignment relationship links units of behavior with active elements (e.g., roles, components) that perform them, or roles with actors that fulfill them.	
Aggregation	The aggregation relationship indicates that an object groups a number of other objects.	
Composition	The composition relationship indicates that an object is composed of one or more other objects.	
Dynamic Relationships		**Notation**
Flow	The flow relationship describes the exchange or transfer of, for example, information or value between processes, function, interactions, and events.	
Triggering	The triggering relationship describes the temporal or causal relationships between processes, functions, interactions, and events.	
Other Relationships		**Notation**
Grouping	The grouping relationship indicates that objects, of the same type or different types, belong together based on some common characteristic.	
Junction	A junction is used to connect relationships of the same type.	
Specialization	The specialization relationship indicates that an object is a specialization of another object.	

7.5 Derived Relationships

The structural relationships described in the previous sections form an important category of relationships to describe coherence. The structural relationships are listed in Table 4 in ascending order by "strength": association is the weakest structural relationship; composition is the strongest. Part of the language definition is an abstraction rule that states that two relationships that join at an intermediate element can be combined and replaced by the weaker of the two.

> If two structural relationships $r{:}R$ and $s{:}S$ are permitted between elements a, b, and c such that $r(a,b)$ and $s(b,c)$, then a structural relationship $t{:}T$ is also permitted, with $t(a,c)$ and type T being the weakest of R and S.

For the application of this rule, it is assumed that the assignment relationship has a direction (as indicated by the role names in Figure 2, Figure 3, Figure 9, Figure 26, Figure 34, and Figure 44).

Transitively applying this property allows us to replace a "chain" of structural relationships (with intermediate model elements) by the weakest structural relationship in the chain. For a more formal description and derivation of this rule we refer to [13].

With this rule, it is possible to determine the "indirect" relationships that exist between model elements without a direct relationship, which may be useful for, among other things, impact analysis. An example is shown in Figure 48: assume that we would like to know what the impact on the client is if the CRM system fails. In this case, an indirect "used by" relationship (the thick arrow on the left) can be derived from this system to the **Claim registration service** (from the chain assignment – used by – realization – used by – realization). No indirect (structural) relationship is drawn between the **CRM system** and the **Claims payment service**.

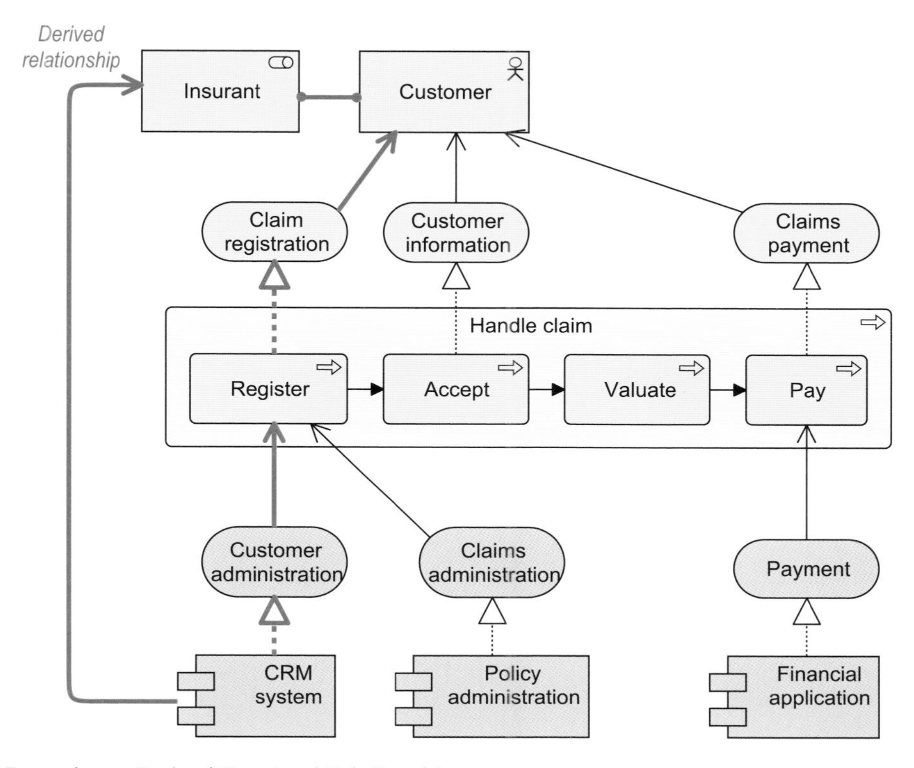

Example 45: Derived Structural Relationship

For the two dynamic relationships, the following rules apply:

- The begin and/or end point of a triggering or flow relationship between behavioral elements (e.g., processes or functions) may be transferred to active structural elements (e.g., business actors or application components) that are assigned to them.
- The begin and/or end point of a triggering or flow relationship between behavior elements may be transferred to services that they realize.

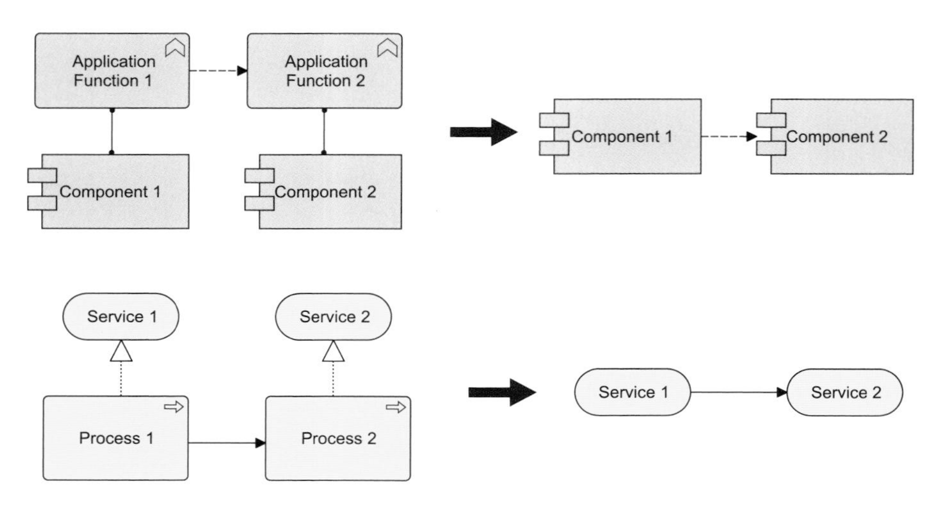

Example 46: Derived Dynamic Relationship

It is important to note that all these derived relationships are also valid in ArchiMate. These are not shown in the "barebones" metamodel illustrations shown in the previous sections, because this would clutter up the diagrams. However, the table in Section E.2 shows all permitted relationships between two elements in the language.

Chapter 8

Architecture Viewpoints

8.1 Introduction

Establishing and maintaining a coherent enterprise architecture is clearly a complex task, because it involves many different people with differing backgrounds using various notations. In order to get a handle on this complexity, researchers have initially focused on the definition of architectural frameworks for classifying and positioning the various architectural descriptions with respect to each other (e.g., the Zachman framework [5], [8]). A problem with looking at enterprise architecture through the lens of an architectural framework is that it categorizes and divides architectural descriptions rather than providing insight into their coherence.

ArchiMate advocates a more flexible approach in which architects and other stakeholders can define their own views on the enterprise architecture. In this approach, views are specified by *viewpoints*. Viewpoints define abstractions on the set of models representing the enterprise architecture, each aimed at a particular type of stakeholder and addressing a particular set of concerns. Viewpoints can both be used to view certain aspects in isolation, and for relating two or more aspects.

The notion of viewpoint-oriented architecture has been around for a while in requirements and software engineering. In the 1990s, a substantial number of researchers worked on what was phrased as "the multiple perspectives problem" [14], [15]. By this term they referred to the problem of how to organize and guide (software) development in a setting with many actors, using diverse representation schemes, having diverse domain knowledge and different development strategies. A general framework has been developed in order to address the diverse issues related to this problem [14], [15]. In this framework, a viewpoint combines the notion of "actor", "role", or "agent" in the development process with the idea of a "perspective" or "view" which an actor maintains. More precisely, viewpoints are defined as loosely coupled, locally managed, distributable objects; thus containing identity, state, and behavior. A viewpoint is more than a "partial specification"; in addition, it contains partial knowledge of how to develop that partial specification. These early ideas on viewpoint-oriented software engineering have found their way into ISO/IEC 42010:2007 [1] on which we have based our definitions below.

As a result of these ideas, several architecture frameworks can be found in the field of literature, which are essentially viewpoint classification schemes. For example, the Zachman framework [5], [8] divides the enterprise architecture into 36 different enterprise-wide "architectures" (i.e., viewpoints). Tapscott and Caston's framework [16] distinguishes five different and complementing viewpoints: business, work, information, application, and technology. Kruchten [17] introduces the "4+1" method, in which four views (logic, process, development, and physical), each having its own notation, are coupled through a fifth view: the scenario view illustrating the collaboration between the other four views.

Viewpoints are also prominently present in the ISO standardized Reference Model for Open Distributed Processing (RM-ODP) [6]. The RM-ODP identifies five viewpoints from which to specify ODP systems, each focusing on a particular area of concern; i.e., enterprise, information, computational, engineering, and technology. It is claimed that the ODP viewpoints form a necessary and sufficient set to meet the needs of ODP standards. More recently, the term "viewpoint" is also used in OMG's Model Driven Architecture (MDA) initiative to refer to the different model types; i.e., Platform-Independent Model (PIM) and Platform-Specific Model (PSM) [18]. Hence, we conclude that the use of viewpoints and architectural views are well-established concepts in software architecture.

In the domain of enterprise architecture, the TOGAF framework describes a taxonomy of views for different categories of stakeholders. Next to this description of views, TOGAF also provides guidelines for the development and use of viewpoints and views in enterprise architecture models.

The views and viewpoints proposed by any of the above mentioned frameworks should not be considered in isolation: views are inter-related and, often, it is exactly a combination of views together with their underlying inter-dependency relationships that is the best way to describe and communicate a piece of architecture. It should, however, be noted that views and viewpoints have a limiting character. They are eventually a restriction of the whole system (and architecture) to a partial number of aspects – a view is just a partial incomplete depiction of the system.

8.2 Views, Viewpoints, and Stakeholders

Views are an ideal mechanism to purposefully convey information about architecture areas. In general, a *view* is defined as a part of an architecture description that addresses a set of related concerns and is addressed to a set of stakeholders. A view is specified by means of a *viewpoint*, which prescribes the concepts, models, analysis techniques, and visualizations that are provided by the view. Simply put, a view is what you see and a viewpoint is where you are looking from.

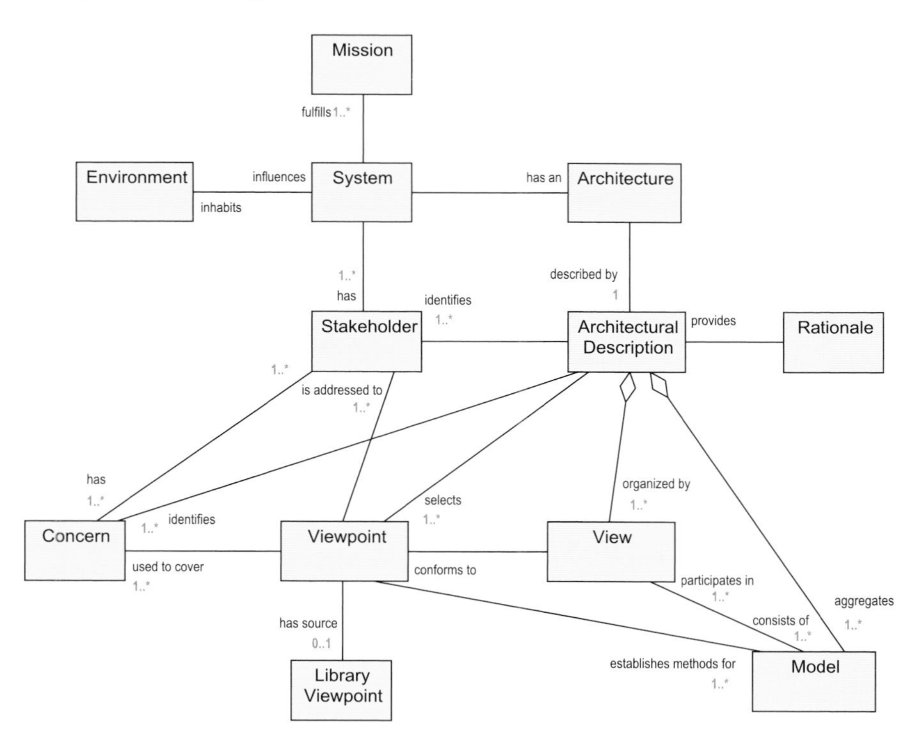

Figure 58: Conceptual Model of Architectural Description (from [1])

Viewpoints are a means to focus on particular aspects of the architecture. These aspects are determined by the concerns of a stakeholder with whom communication takes place. What should and should not be visible from a specific viewpoint is therefore entirely dependent on the argumentation with respect to a stakeholder's concerns.

Viewpoints are designed for the purpose of communicating certain aspects of an architecture. The communication enabled by a viewpoint can be strictly informative, but in general is bi-directional. The architect informs stakeholders, and stakeholders give their feedback (critique or consent) on the presented aspects. What is and what is not shown in a view depends on the scope of the viewpoint and on what is relevant to the concerns of the stakeholder. Ideally, these are the same; i.e., the viewpoint is designed with specific concerns of a stakeholder in mind. Relevance to a stakeholder's concern, therefore, is *the* selection criterion that is used to determine which objects and relationships are to appear in a view.

The following are examples of stakeholders and concerns as a basis for the specification of viewpoints:

- *End user*: For example, what are the consequences for his work and workplace?
- *Architect*: What is the consequence for the maintainability of a system, with respect to corrective, preventive, and adaptive maintenance?
- *Upper-level management*: How can we ensure our policies are followed in the development and operation of processes and systems? What is the impact of decisions (on personnel, finance, ICT, etc.)?
- *Operational manager*: Responsible for exploitation or maintenance: For example, what new technologies are there to prepare for? Is there a need to adapt maintenance processes? What is the impact of changes to existing applications? How secure are my systems?
- *Project manager*: Responsible for the development of new applications: What are the relevant domains and their relationships? What is the dependence of business processes on the applications to be built? What is their expected performance?
- *Developer*: What are the modifications with respect to the current situation that need to be done?

8.3 Viewpoint Classification

An architect is confronted with many different types of stakeholders and concerns. To help him in selecting the right viewpoints for the task at hand, we introduce a framework for the definition and classification of viewpoints and views. The framework is based on two dimensions: *purpose* and *content*. The following three types of architecture support the purpose dimension of architecture views:

- *Designing*: Design viewpoints support architects and designers in the design process from initial sketch to detailed design. Typically, design viewpoints consist of diagrams, like those used in, for example, UML.
- *Deciding*: Decision support viewpoints assist managers in the process of decision-making by offering insight into cross-domain architecture relationships, typically through projections and intersections of underlying models, but also by means of analytical techniques. Typical examples are cross-reference tables, landscape maps, lists, and reports.
- *Informing*: Informing viewpoints help to inform any stakeholder about the enterprise architecture, in order to achieve understanding, obtain commitment, and convince adversaries. Typical examples are illustrations, animations, cartoons, flyers, etc.

The goal of this classification is to assist architects and others find suitable viewpoints given their task at hand; i.e., the purpose that a view must serve and the content it should display. With the help of this framework, it is easier to find typical viewpoints that might be useful in a given situation. This implies that we do not provide an orthogonal categorization of each viewpoint into one of three classes; these categories are not exclusive in the sense that a viewpoint in one category cannot be applied to achieve another type of support. For instance, some decision support viewpoints may be used to communicate to any other stakeholders as well.

For characterizing the content of a view we define the following abstraction levels:

- *Details*: Views on the detailed level typically consider one layer and one aspect from the ArchiMate framework. Typical stakeholders are a software engineer responsible for design and implementation of a software component or a process owner responsible for effective and efficient process execution. Examples of views are a BPMN process diagram and a UML class diagram.
- *Coherence*: At the coherence abstraction level, multiple layers or multiple aspects are spanned. Extending the view to more than one layer or aspect enables the stakeholder to focus on architecture relationships like process-

uses-system (multiple layer) or application-uses-object (multiple aspect). Typical stakeholders are operational managers responsible for a collection of IT services or business processes.

- *Overview*: The overview abstraction level addresses both multiple layers and multiple aspects. Typically, such overviews are addressed to enterprise architects and decision-makers, such as CEOs and CIOs.

In Figure 59, the dimensions of purpose and abstraction level are visualized in a single picture, together with examples of typical stakeholders that are addressed by these viewpoints. The top half of this figure shows the purpose dimension, while the bottom half shows the level of abstraction (or detail). Table 5 and Table 6 summarize the different purposes and abstraction levels.

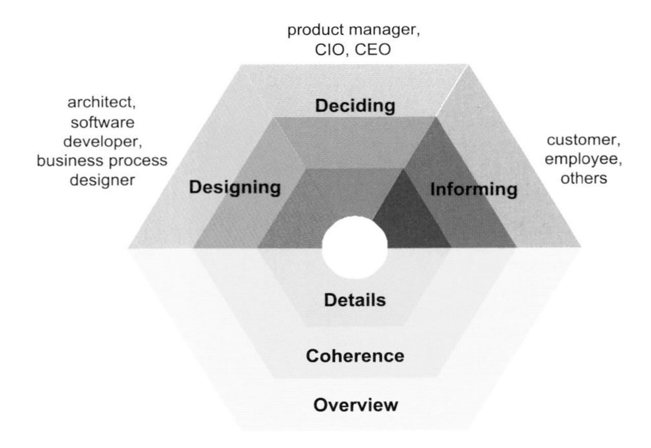

Figure 59: Classification of Enterprise Architecture Viewpoints

Table 5: Viewpoint Purpose

	Typical Stakeholders	Purpose	Examples
Designing	architect, software developer, business process designer	navigate, design, support design decisions, compare alternatives	UML diagram, BPMN diagram, flowchart, ER diagram
Deciding	manager, CIO, CEO	decision-making	cross-reference table, landscape map, list, report
Informing	employee, customer, others	explain, convince, obtain commitment	animation, cartoon, process illustration, chart

Table 6: Viewpoint Abstraction Levels

	Typical Stakeholders	Purpose	Examples
Details	software engineer, process owner	design, manage	UML class diagram, BPMN process diagram
Coherence	operational managers	analyze dependencies, impact of-change	views expressing relationships like "use", "realize", and "assign"
Overview	enterprise architect, CIO, CEO	change management	landscape map

8.4 Standard Viewpoints in ArchiMate

A viewpoint in ArchiMate is a selection of a relevant subset of the ArchiMate concepts (and their relationships) and the representation of that part of an architecture that is expressed in different diagrams. A set of such viewpoints was developed based on practical experience. Some of these viewpoints have a scope that is limited to a single layer or aspect. Thus, the Business Function and Business Process viewpoints show the two main perspectives on the business behavior; the Organization viewpoint depicts the structure of the enterprise in terms of its departments, roles, etc.; the Information Structure viewpoint describes the information and data used; the Application Structure, Behavior, and Co-operation viewpoints contain the applications and components and their mutual relationships; and the Infrastructure viewpoint shows the infrastructure and platforms underlying the enterprise's information systems in terms of networks, devices, and system software. Other viewpoints link multiple layers and/or aspects: the Actor Co-operation and Product viewpoints relate the enterprise to its environment; the Application Usage viewpoint relates applications to their use in, for example, business processes; and the Deployment viewpoint shows how applications are mapped onto the underlying infrastructure.

In the following sections, the ArchiMate viewpoints are described in detail. For each viewpoint the comprised concepts and relationships, the guidelines for the viewpoint use, and the goal and target group and of the viewpoint are indicated. Furthermore, each viewpoint description contains example models. For more details on the goal and use of viewpoints, refer to [2],

Chapter 8. The diagrams illustrating the permitted concepts and relationships for each viewpoint do not show all permitted relationships: every element in a given viewpoint can have composition, aggregation, and specialization relationships with elements of the same type; furthermore, there are indirect relationships that can be derived as explained in Section 6.5.

8.4.1 Introductory Viewpoint

The Introductory viewpoint forms a subset of the full ArchiMate language using a simplified notation. It is typically used at the start of a design trajectory, when not everything needs to be detailed yet, or to explain the essence of an architecture model to non-architects that require a simpler, more intuitive notation. Another use of this basic, less formal viewpoint is that it tries to avoid the impression that the architectural design is already fixed, an idea that may easily arise when using a more formal, highly structured or detailed visualization.

We use a simplified notation for the concepts (e.g., a cloud to represent a network, as is common in informal diagrams of the technical infrastructure), and for the relationships. All relationships except "triggering" and "realization" are denoted by simple lines; "realization" has an arrow in the direction of the realized service; "triggering" is also represented by an arrow. The concepts are denoted with slightly thicker lines and rounded corners, which give a less formal impression. The example below illustrates this notation. On purpose, the layout of this example is not as "straight" as an ordinary architecture diagram; this serves to avoid the idea that the design is already fixed.

Table 7: Introductory Viewpoint Description

Introductory Viewpoint	
Stakeholders	Enterprise architects, managers
Concerns	Make design choices visible, convince stakeholders
Purpose	Designing, deciding, informing
Abstraction Level	Coherence, Overview, Detail
Layer	Business, Application, and Technology layers (see also Figure 4)
Aspects	Structure, behavior, information (see also Figure 4)

Concepts and Relationships

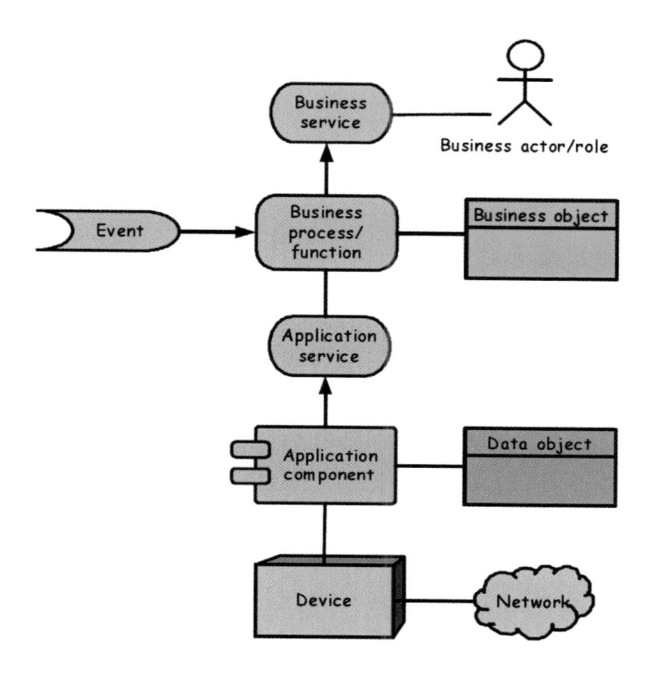

Example

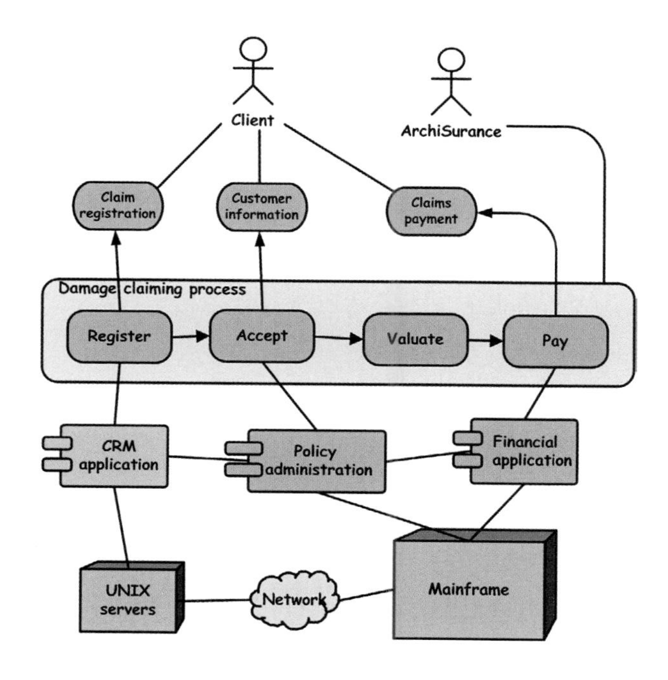

8.4.2 Organization Viewpoint

The Organization viewpoint focuses on the (internal) organization of a company, a department, a network of companies, or of another organizational entity. It is possible to present models in this viewpoint as nested block diagrams, but also in a more traditional way, such as organizational charts. The Organization viewpoint is very useful in identifying competencies, authority, and responsibilities in an organization.

Table 8: Organization Viewpoint Description

Organization Viewpoint	
Stakeholders	Enterprise, process and domain architects, managers, employees, shareholders
Concerns	Identification of competencies, authority, and responsibilities
Purpose	Designing, deciding, informing
Abstraction Level	Coherence
Layer	Business layer (see also Figure 4)
Aspects	Structure (see also Figure 4)

Concepts and Relationships

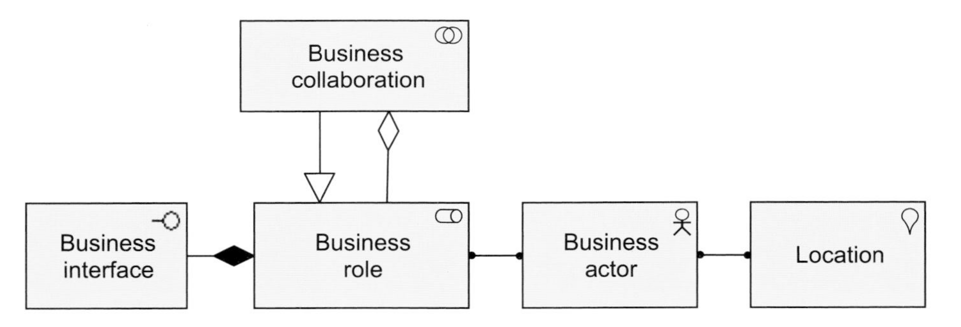

Example

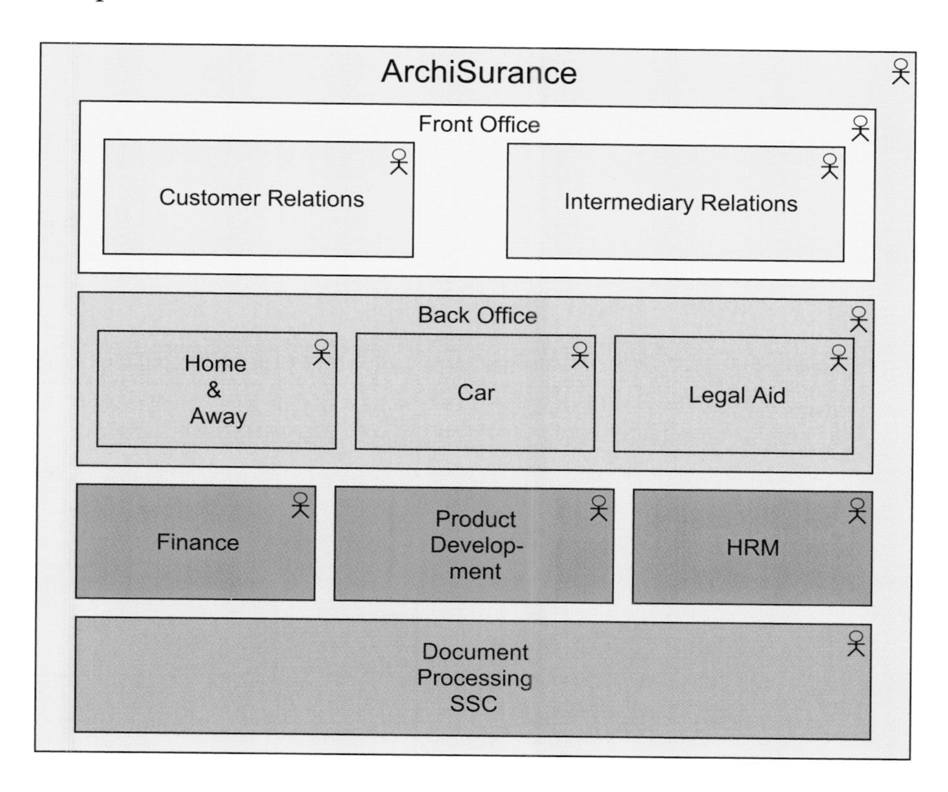

8.4.3 Actor Co-operation Viewpoint

The Actor Co-operation viewpoint focuses on the relationships of actors with each other and their environment. A common example of this is the "context diagram", which puts an organization into its environment, consisting of external parties such as customers, suppliers, and other business partners. It is very useful in determining external dependencies and collaborations and shows the value chain or network in which the actor operates.

Another important use of the Actor Co-operation viewpoint is in showing how a number of co-operating business actors and/or application components together realize a business process. Hence, in this view, both business actors or roles and application components may occur.

Table 9: Actor Co-operation Viewpoint Description

Actor Co-operation Viewpoint		
Stakeholders	Enterprise, process, and domain architects	
Concerns	Relationships of actors with their environment	
Purpose	Designing, deciding, informing	
Abstraction Level	Detail	
Layer	Business layer (application layer) (see also Figure 4)	
Aspects	Structure, behavior (see also Figure 4)	

Concepts and Relationships

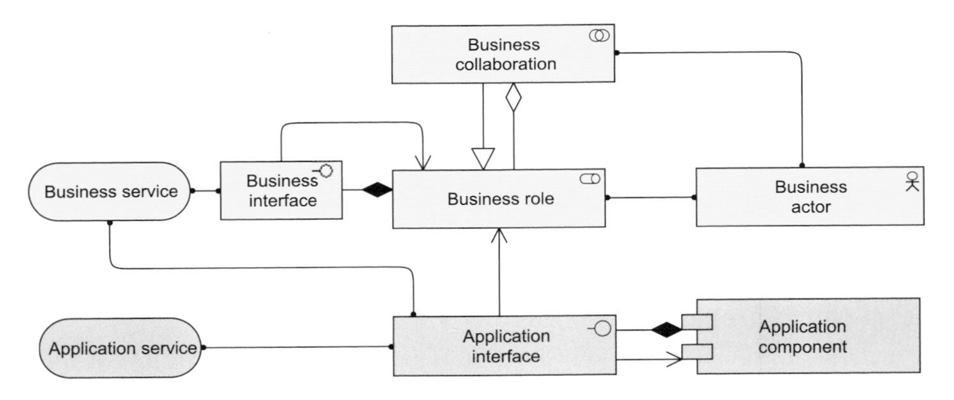

Example

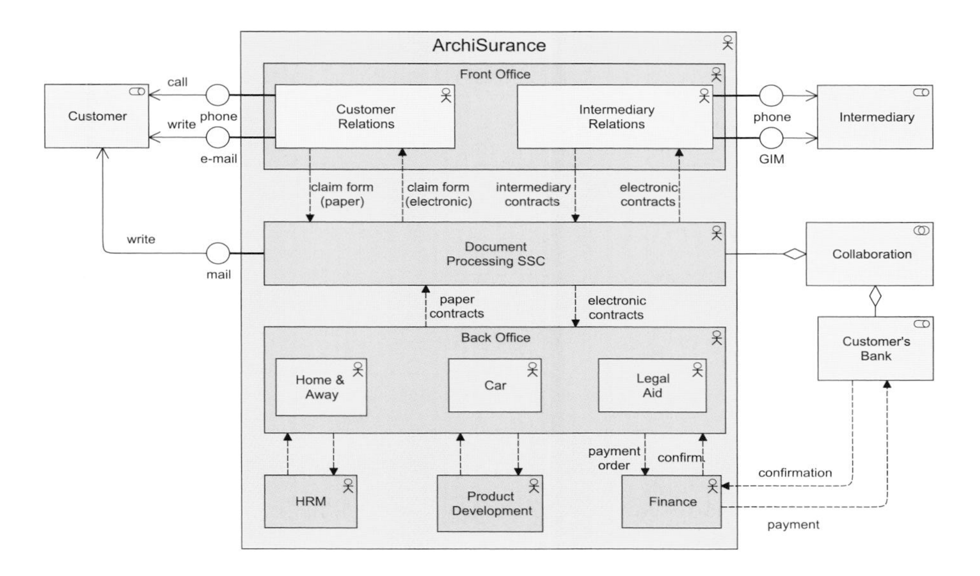

8.4.4 Business Function Viewpoint

The Business Function viewpoint shows the main business functions of an organization and their relationships in terms of the flows of information, value, or goods between them. Business functions are used to represent the most stable aspects of a company in terms of the primary activities it performs, regardless of organizational changes or technological developments. Therefore, the business function architecture of companies that operate in the same market often exhibit close similarities. The business function viewpoint thus provides high-level insight in the general operations of the company, and can be used to identify necessary competencies, or to structure an organization according to its main activities.

Table 10: Business Function Viewpoint Description

Business Function Viewpoint	
Stakeholders	Enterprise, process, and domain architects
Concerns	Identification of competencies, identification of main activities, reduction of complexity
Purpose	Designing
Abstraction Level	Coherence
Layer	Business layer (see also Figure 4)
Aspects	Behavior, structure (see also Figure 4)

Concepts and Relationships

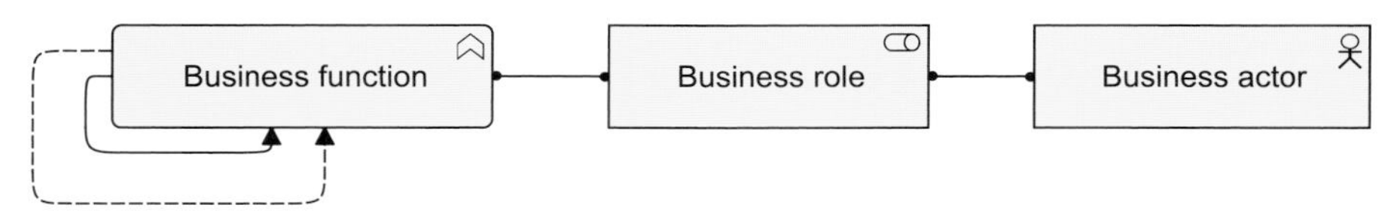

Example

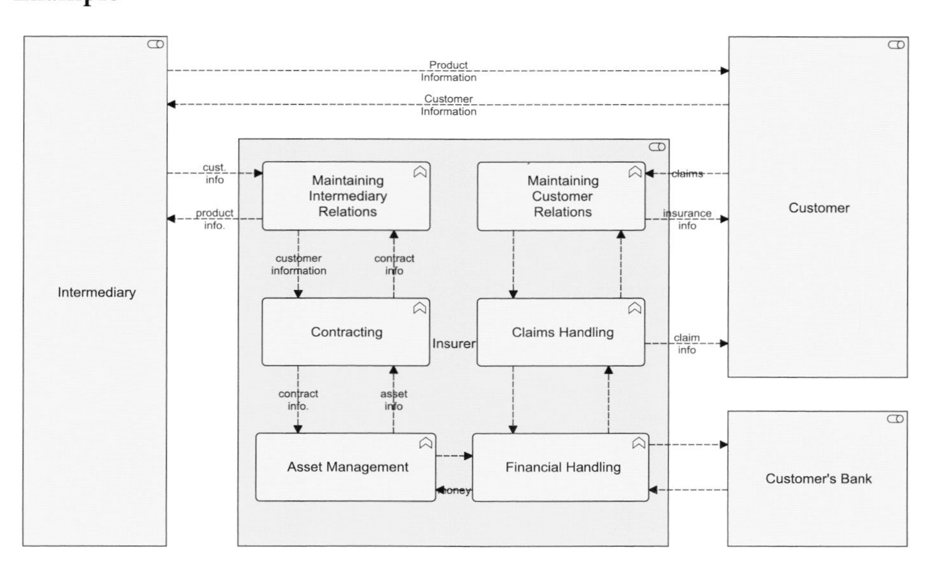

8.4.5 Business Process Viewpoint

The Business Process viewpoint is used to show the high-level structure and composition of one or more business processes. Next to the processes themselves, this viewpoint contains other directly related concepts, such as:

- The services that a business process offers to the outside world, showing how a process contributes to the realization of the company's products
- The assignment of business processes to roles, which gives insight into the responsibilities of the associated actors
- The information used by the business process

Each of these can be regarded as a "sub-view" of the business process view.

Table 11: Business Process Viewpoint Description

Business Process Viewpoint	
Stakeholders	Process and domain architects, operational managers
Concerns	Structure of business processes, consistency and completeness, responsibilities
Purpose	Designing
Abstraction Level	Detail
Layer	Business layer (see also Figure 4)
Aspects	Behavior (see also Figure 4)

Concepts and Relationships

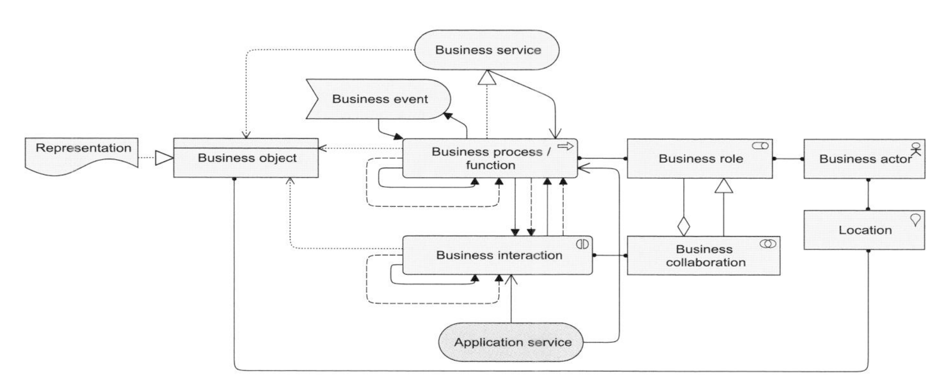

Example

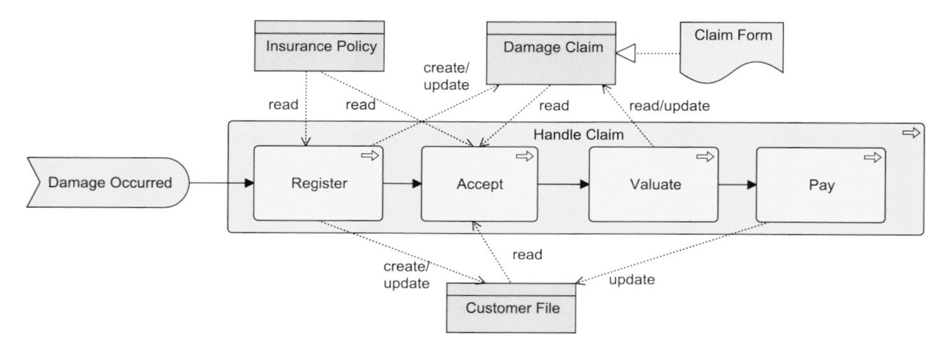

8.4.6 Business Process Co-operation Viewpoint

The Business Process Co-operation viewpoint is used to show the relationships of one or more business processes with each other and/or with their environment. It can both be used to create a high-level design of business processes within their context and to provide an operational manager responsible for one or more such processes with insight into their dependencies. Important aspects of business process co-operation are:

- Causal relationships between the main business processes of the enterprise
- Mapping of business processes onto business functions
- Realization of services by business processes
- Use of shared data

Each of these can be regarded as a "sub-view" of the business process co-operation view.

Table 12: Business Process Co-operation Viewpoint Description

Business Process Co-operation Viewpoint	
Stakeholders	Process and domain architects, operational managers
Concerns	Dependencies between business processes, consistency and completeness, responsibilities
Purpose	Designing, deciding
Abstraction Level	Coherence
Layer	Business layer, application layer (see also Figure 4)
Aspects	Behavior (see also Figure 4)

Concepts and Relationships

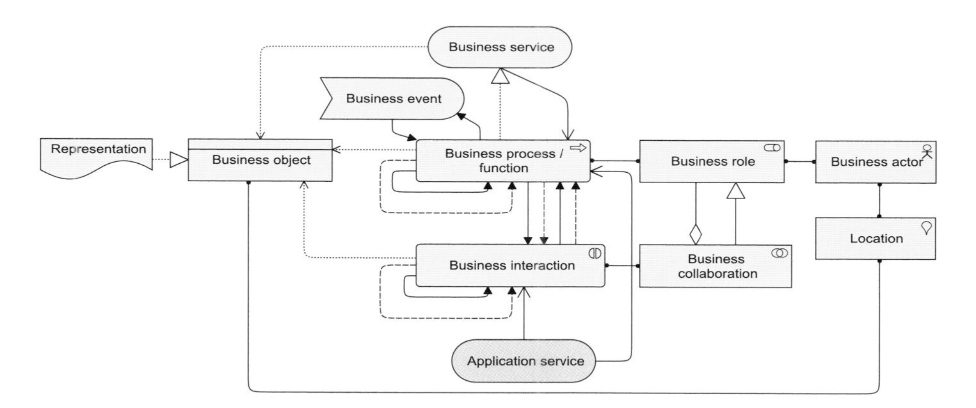

Example

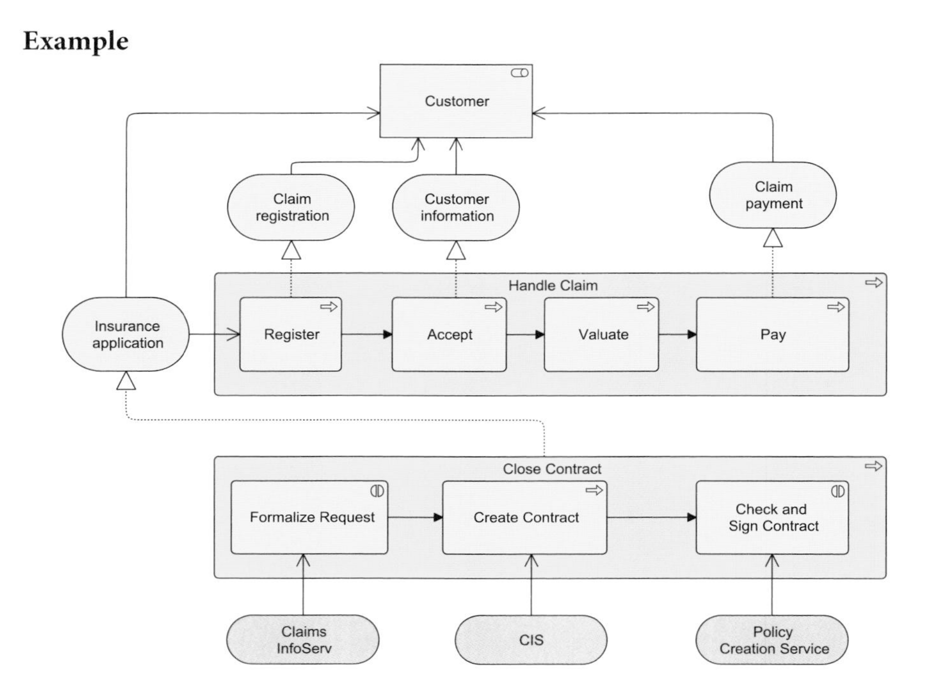

8.4.7 Product Viewpoint

The Product viewpoint depicts the value that these products offer to the customers or other external parties involved and shows the composition of one or more products in terms of the constituting (business or application) services, and the associated contract(s) or other agreements. It may also be used to show the interfaces (channels) through which this product is offered, and the events associated with the product. A Product viewpoint is typically used in product development to design a product by composing existing services or by identifying which new services have to be created for this product, given the value a customer expects from it. It may then serve as input for business process architects and others that need to design the processes and ICT realizing these products.

Table 13: Product Viewpoint Description

Product Viewpoint	
Stakeholders	Product developers, product managers, process and domain architects
Concerns	Product development, value offered by the products of the enterprise
Purpose	Designing, deciding
Abstraction Level	Coherence
Layer	Business layer, application layer (see also Figure 4)
Aspects	Behavior, information (see also Figure 4)

Concepts and Relationships

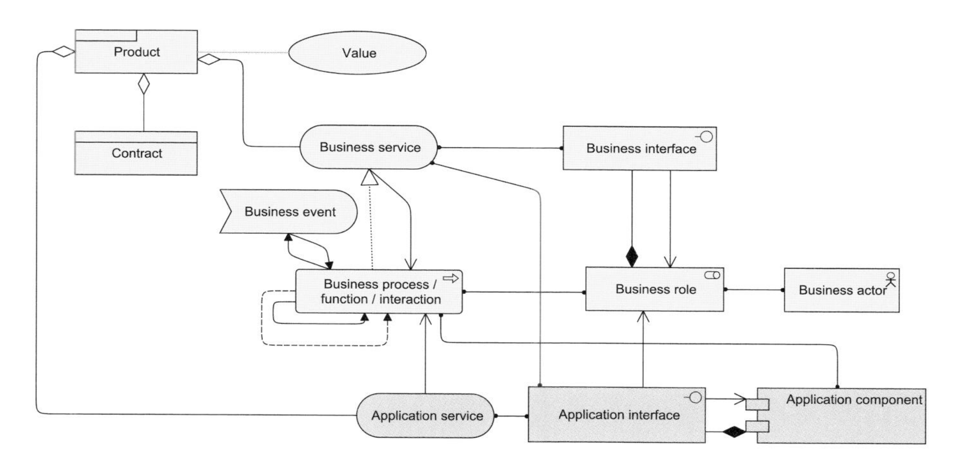

Example

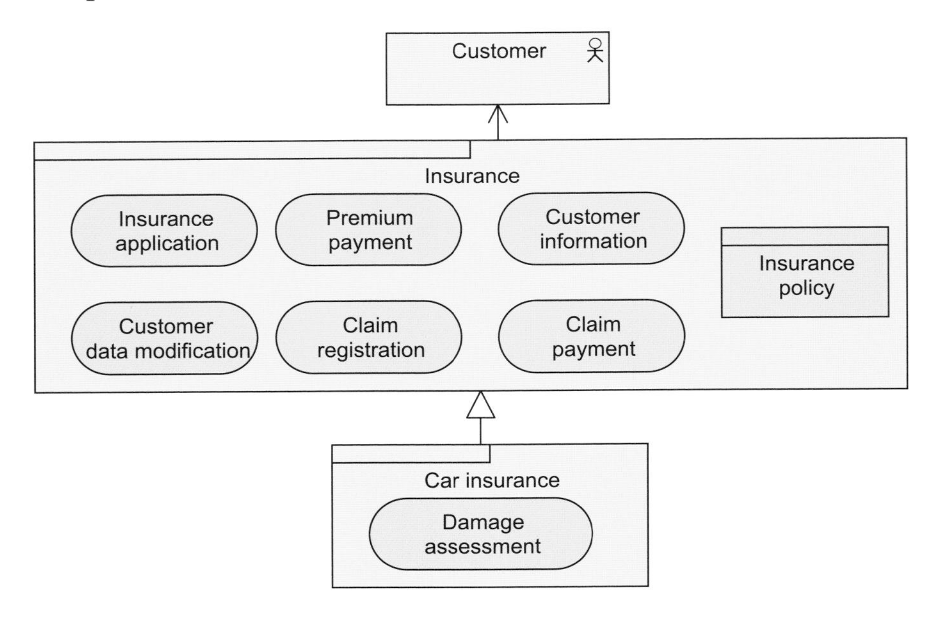

8.4.8 Application Behavior Viewpoint

The Application Behavior viewpoint describes the internal behavior of an application; e.g., as it realizes one or more application services. This viewpoint is useful in designing the main behavior of applications, or in identifying functional overlap between different applications.

Table 14: Application Behavior Viewpoint Description

Application Behavior Viewpoint		
Stakeholders	Enterprise, process, application, and domain architects	
Concerns	Structure, relationships and dependencies between applications, consistency and completeness, reduction of complexity	
Purpose	Designing	
Abstraction Level	Coherence, details	
Layer	Application layer (see also Figure 4)	
Aspects	Information, behavior, structure (see also Figure 4)	

Concepts and Relationships

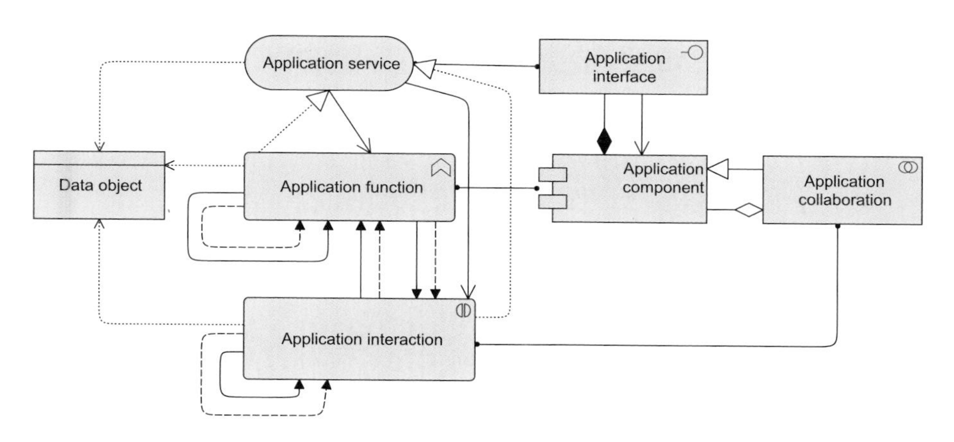

Example

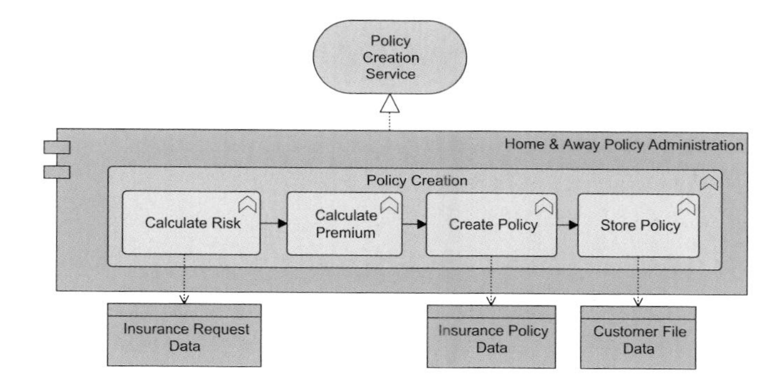

8.4.9 Application Co-operation Viewpoint

The Application Co-operation viewpoint describes the relationships between applications components in terms of the information flows between them, or in terms of the services they offer and use. This viewpoint is typically used to create an overview of the application landscape of an organization. This viewpoint is also used to express the (internal) co-operation or orchestration of services that together support the execution of a business process.

Table 15: Application Co-operation Viewpoint Description

Application Co-operation Viewpoint		
Stakeholders	Enterprise , process, application, and domain architects	
Concerns	Relationships and dependencies between applications, orchestration/choreography of services, consistency and completeness, reduction of complexity	
Purpose	Designing	
Abstraction Level	Coherence, details	
Layer	Application layer (see also Figure 4)	
Aspects	Behavior, structure (see also Figure 4)	

Concepts and Relationships

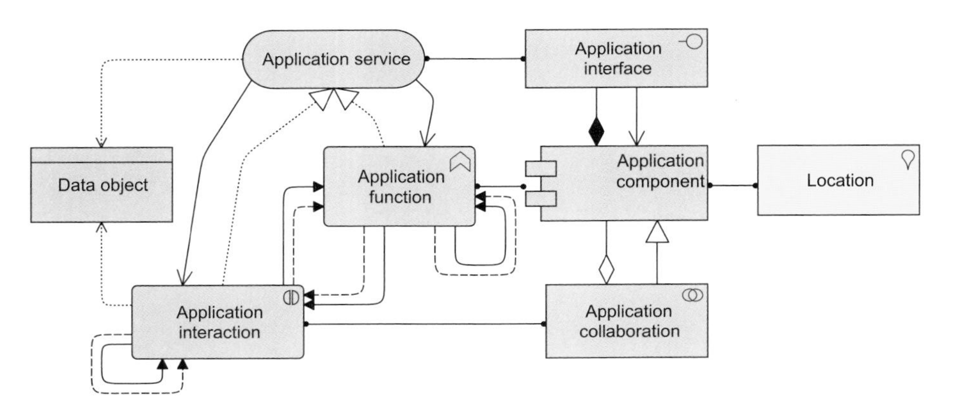

Example

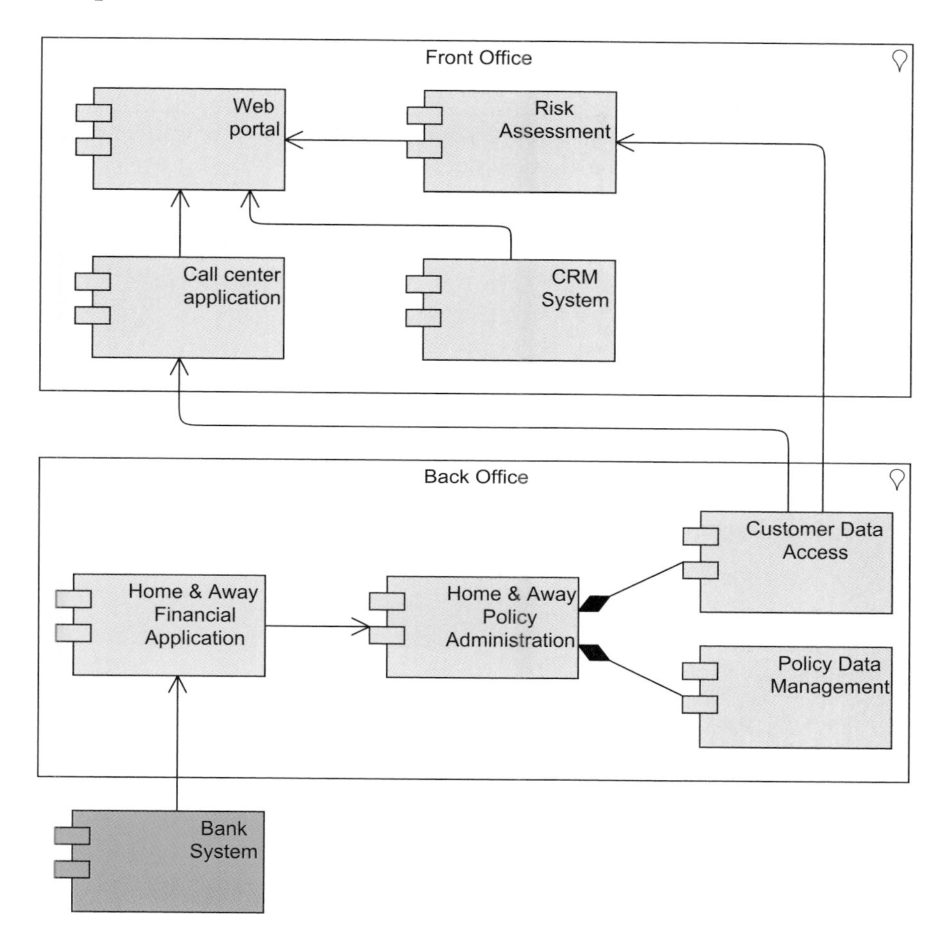

8.4.10 Application Structure Viewpoint

The Application Structure viewpoint shows the structure of one or more applications or components. This viewpoint is useful in designing or understanding the main structure of applications or components and the associated data; e.g., to break down the structure of the system under construction, or to identify legacy application components that are suitable for migration/integration.

Table 16: Application Structure Viewpoint Description

Application Structure Viewpoint		
Stakeholders	Enterprise, process, application, and domain architects	
Concerns	Application structure, consistency and completeness, reduction of complexity	
Purpose	Designing	
Abstraction Level	Details	
Layer	Application layer (see also Figure 4)	
Aspects	Structure, information (see also Figure 4)	

Concepts and Relationships

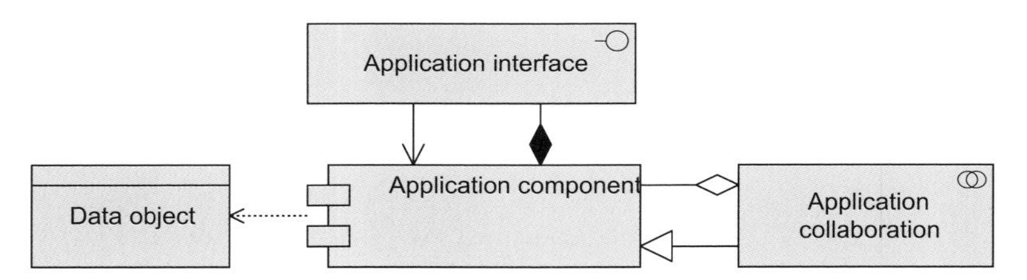

Example

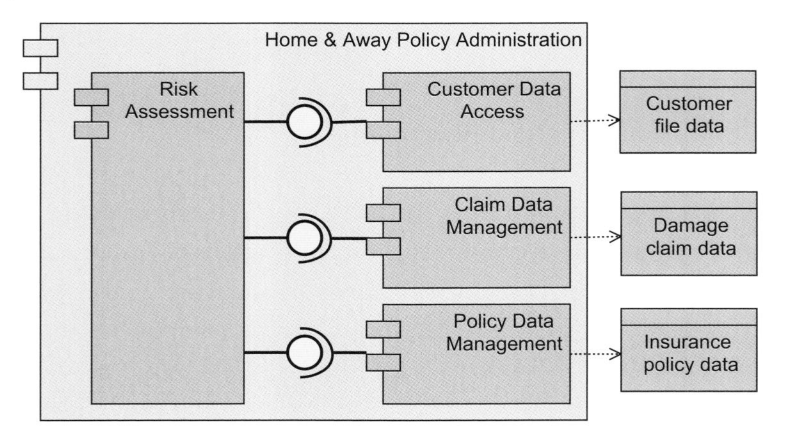

8.4.11 Application Usage Viewpoint

The Application Usage viewpoint describes how applications are used to support one or more business processes, and how they are used by other applications. It can be used in designing an application by identifying the services needed by business processes and other applications, or in designing business processes by describing the services that are available. Furthermore, since it identifies the dependencies of business processes upon applications, it may be useful to operational managers responsible for these processes.

Table 17: Application Usage Viewpoint Description

Application Usage Viewpoint	
Stakeholders	Enterprise, process, and application architects, operational managers
Concerns	Consistency and completeness, reduction of complexity
Purpose	Designing, deciding
Abstraction Level	Coherence
Layer	Business and application layers (see also Figure 4)
Aspects	Behavior, structure (see also Figure 4)

Concepts and Relationships

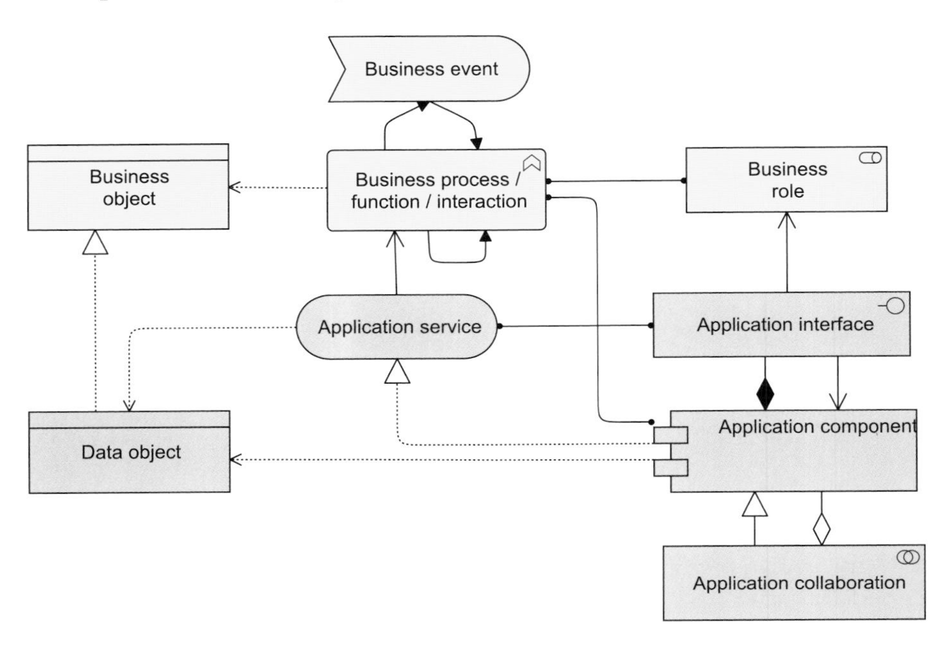

Example

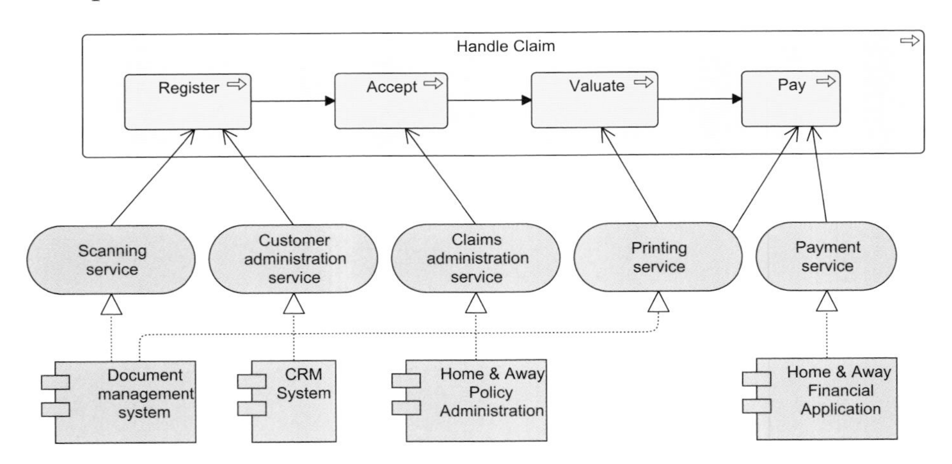

8.4.12 Infrastructure Viewpoint

The Infrastructure viewpoint contains the software and hardware infrastructure elements supporting the application layer, such as physical devices, networks, or system software (e.g., operating systems, databases, and middleware).

Table 18: Infrastructure Viewpoint Description

Infrastructure Viewpoint	
Stakeholders	Infrastructure architects, operational managers
Concerns	Stability, security, dependencies, costs of the infrastructure
Purpose	Designing
Abstraction Level	Details
Layer	Technology layer (see also Figure 4)
Aspects	Behavior, structure (see also Figure 4)

Concepts and Relationships

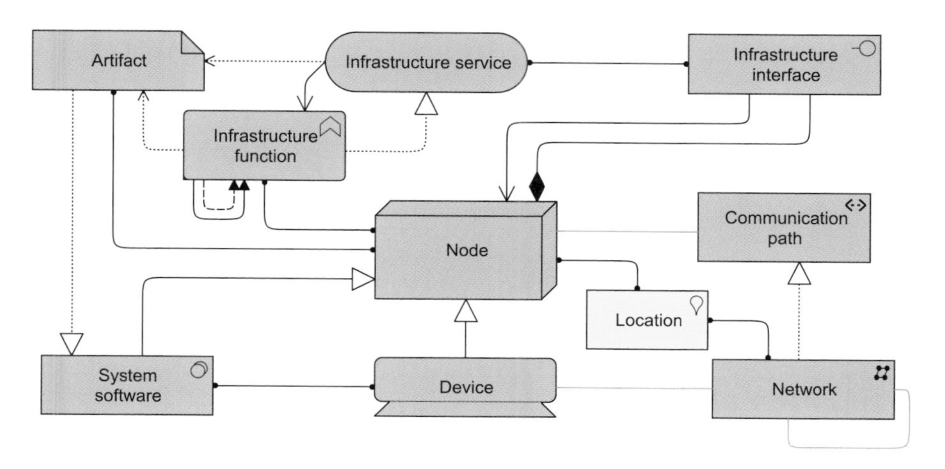

Example

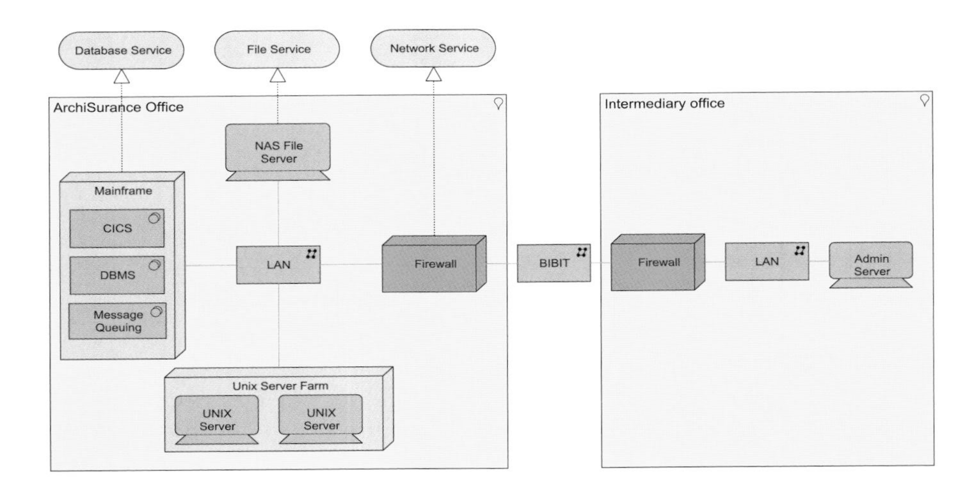

8.4.13 Infrastructure Usage Viewpoint

The Infrastructure Usage viewpoint shows how applications are supported by the software and hardware infrastructure: the infrastructure services are delivered by the devices; system software and networks are provided to the applications. This viewpoint plays an important role in the analysis of performance and scalability, since it relates the physical infrastructure to the logical world of applications. It is very useful in determining the performance and quality requirements on the infrastructure based on the demands of the various applications that use it.

Table 19: Infrastructure Usage Viewpoint Description

Infrastructure Usage Viewpoint		
Stakeholders	Application, infrastructure architects, operational managers	
Concerns	Dependencies, performance, scalability	
Purpose	Designing	
Abstraction Level	Coherence	
Layer	Application and technology layers (see also Figure 4)	
Aspects	Behavior, structure (see also Figure 4)	

Concepts and Relationships

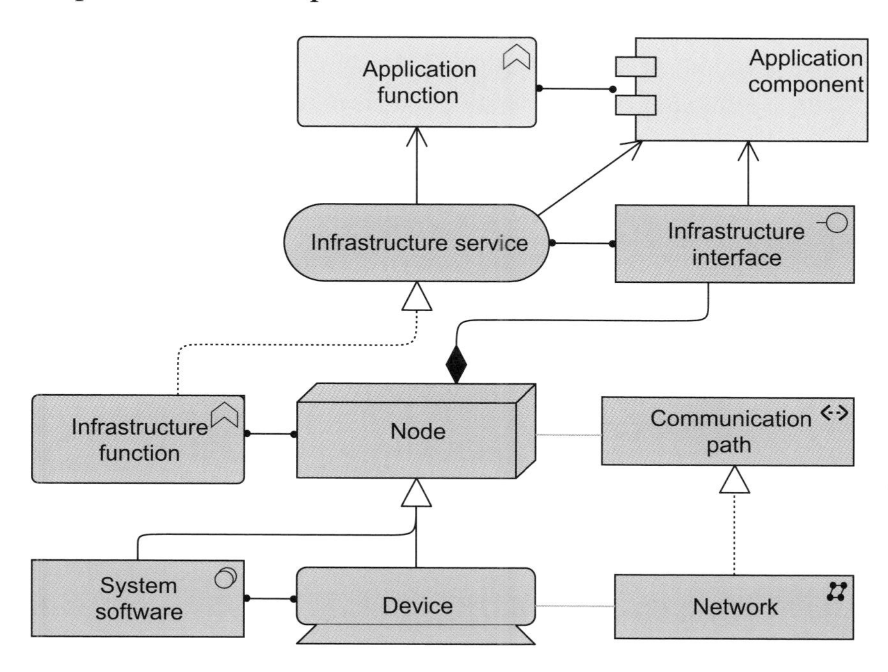

Example

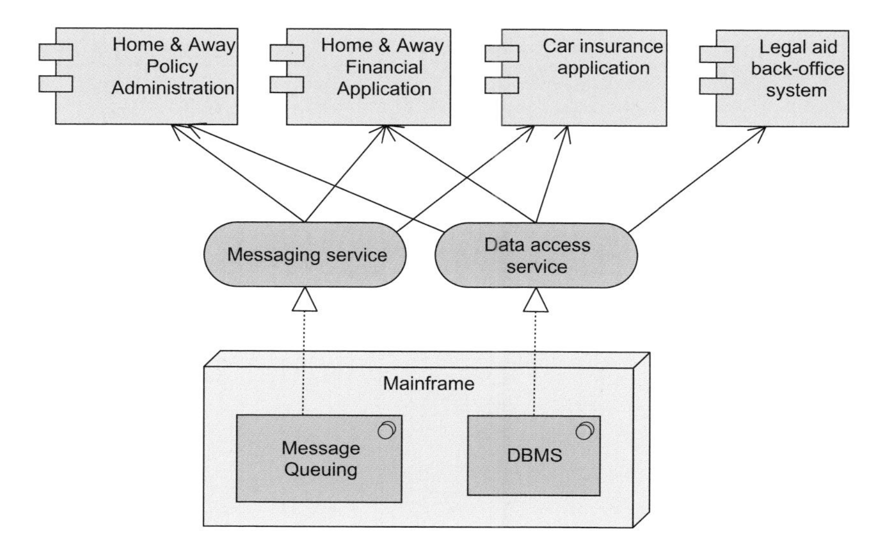

8.4.14 Implementation and Deployment Viewpoint

The Implementation and Deployment viewpoint shows how one or more applications are realized on the infrastructure. This comprises the mapping of (logical) applications and components onto (physical) artifacts, such as Enterprise Java Beans, and the mapping of the information used by these applications and components onto the underlying storage infrastructure; e.g., database tables or other files. Deployment views play an important role in the analysis of performance and scalability, since they relate the physical infrastructure to the logical world of applications. In security and risk analysis, deployment views are used to identify, for example, critical dependencies and risks.

Table 20: Implementation and Deployment Viewpoint Description

Implementation and Deployment Viewpoint		
Stakeholders	Application and infrastructure architects, operational managers	
Concerns	Dependencies, security, risks	
Purpose	Designing	
Abstraction Level	Coherence	
Layer	Application layer, technology layer (see also Figure 4)	
Aspects	Information, behavior, structure (see also Figure 4)	

Concepts and Relationships

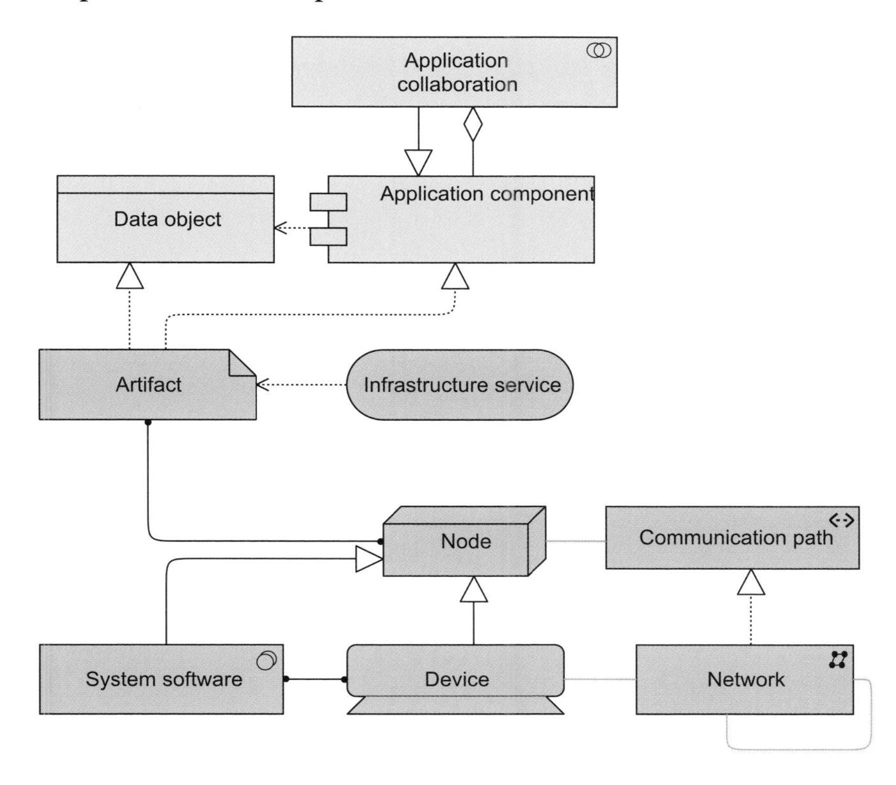

Example

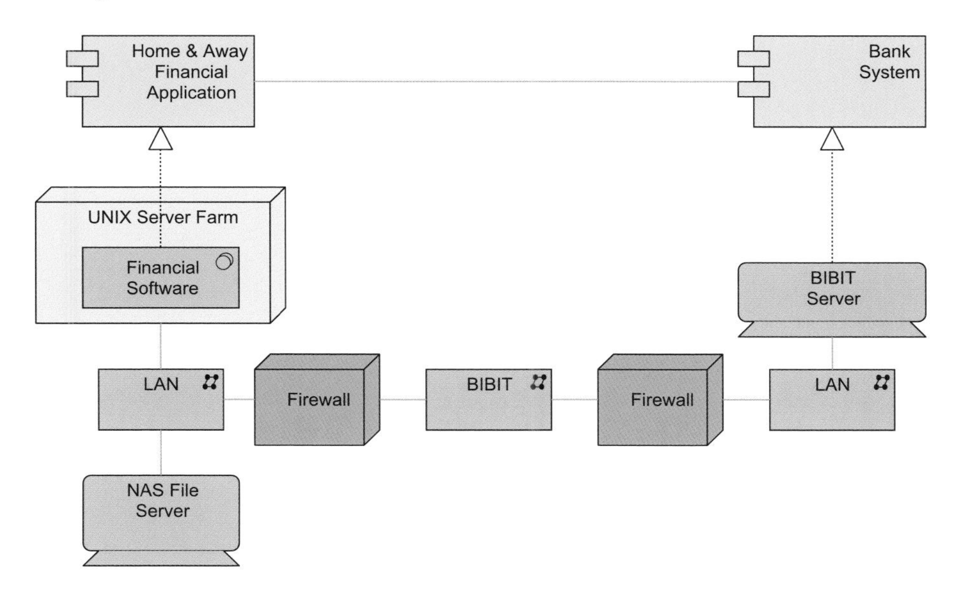

8.4.15 Information Structure Viewpoint

The Information Structure viewpoint is comparable to the traditional information models created in the development of almost any information system. It shows the structure of the information used in the enterprise or in a specific business process or application, in terms of data types or (object-oriented) class structures. Furthermore, it may show how the information at the business level is represented at the application level in the form of the data structures used there, and how these are then mapped onto the underlying infrastructure; e.g., by means of a database schema.

Table 21: Information Structure Viewpoint Description

Information Structure Viewpoint		
Stakeholders	Domain and information architects	
Concerns	Structure and dependencies of the used data and information, consistency and completeness	
Purpose	Designing	
Abstraction Level	Details	
Layer	Business layer, application layer, technology layer (see also Figure 4)	
Aspects	Information (see also Figure 4)	

Concepts and Relationships

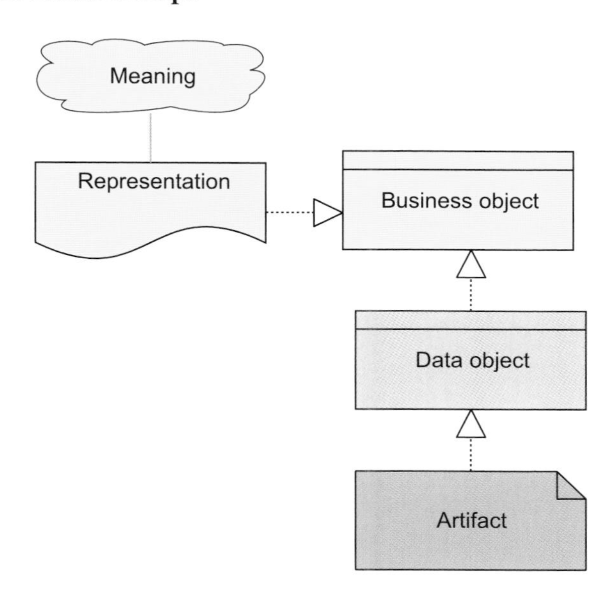

Example

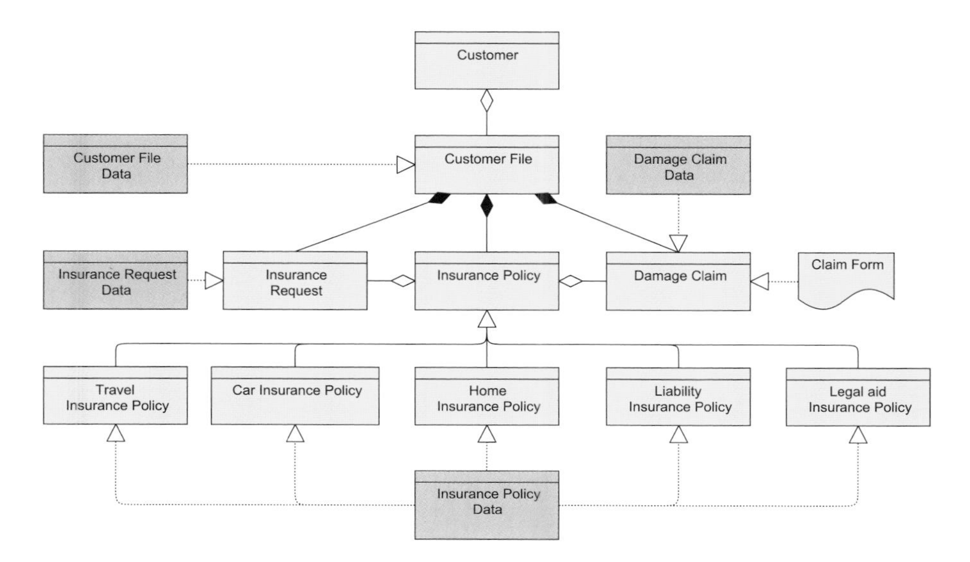

8.4.16 Service Realization Viewpoint

The Service Realization viewpoint is used to show how one or more business services are realized by the underlying processes (and sometimes by application components). Thus, it forms the bridge between the business products viewpoint and the business process view. It provides a "view from the outside" on one or more business processes.

Table 22: Service Realization Viewpoint Description

Service Realization Viewpoint	
Stakeholders	Process and domain architects, product and operational managers
Concerns	Added-value of business processes, consistency and completeness, responsibilities
Purpose	Designing, deciding
Abstraction Level	Coherence
Layer	Business layer (application layer) (see also Figure 4)
Aspects	Behavior, structure, information (see also Figure 4)

Concepts and Relationships

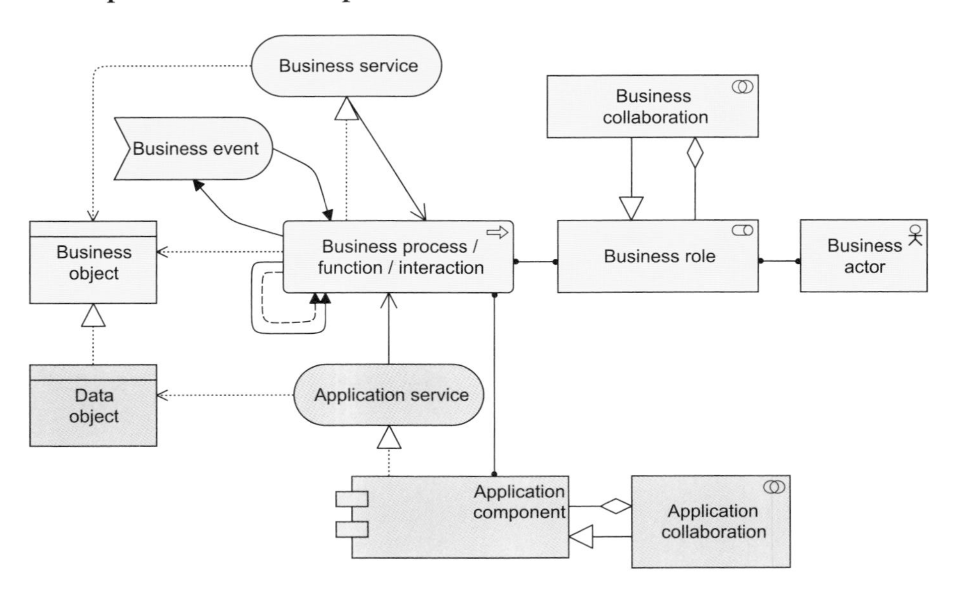

Example

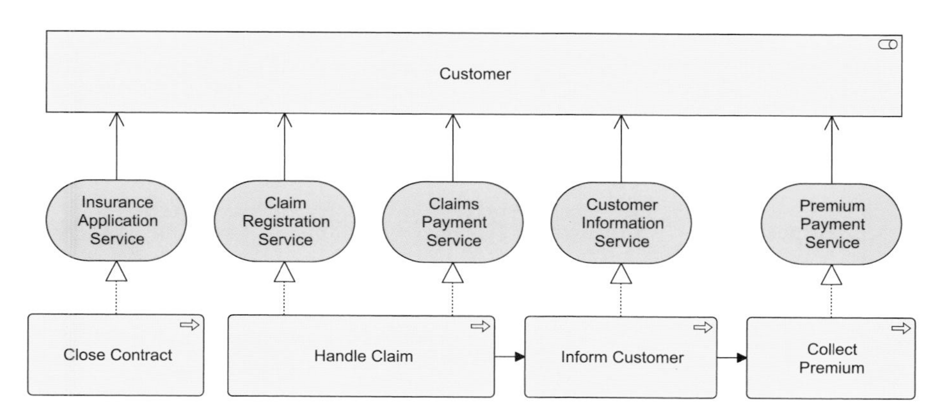

8.4.17 Layered Viewpoint

The Layered viewpoint pictures several layers and aspects of an enterprise architecture in one diagram. There are two categories of layers, namely *dedicated layers* and *service layers*. The layers are the result of the use of the "grouping" relationship for a natural partitioning of the entire set of objects and relationships that belong to a model. The infrastructure, the application, the process, and the actors/roles layers belong to the first category. The structural principle behind a fully layered viewpoint is that each dedicated layer exposes, by means of the "realization" relationship, a layer of services, which are further on "used by" the next dedicated layer. Thus, we can easily separate the internal structure and organization of a dedicated layer from its externally observable behavior expressed as the service layer that the dedicated layer realizes. The order, number, or nature of these layers are not fixed, but in general a (more or less) complete and natural layering of an ArchiMate model should contain the succession of layers depicted in the example given below. However, this example is by no means intended to be prescriptive. The main goal of the Layered viewpoint is to provide overview in one diagram. Furthermore, this viewpoint can be used as support for impact of change analysis and performance analysis or for extending the service portfolio.

Table 23: Layered Viewpoint Description

Layered Viewpoint		
Stakeholders	Enterprise, process, application, infrastructure, and domain architects	
Concerns	Consistency, reduction of complexity, impact of change, flexibility	
Purpose	Designing, deciding, informing	
Abstraction Level	Overview	
Layer	Business layer, application layer, technology layer (see also Figure 4)	
Aspects	Information, behavior, structure (see also Figure 4)	

Concepts and Relationships

All concepts and all relationships.

Example

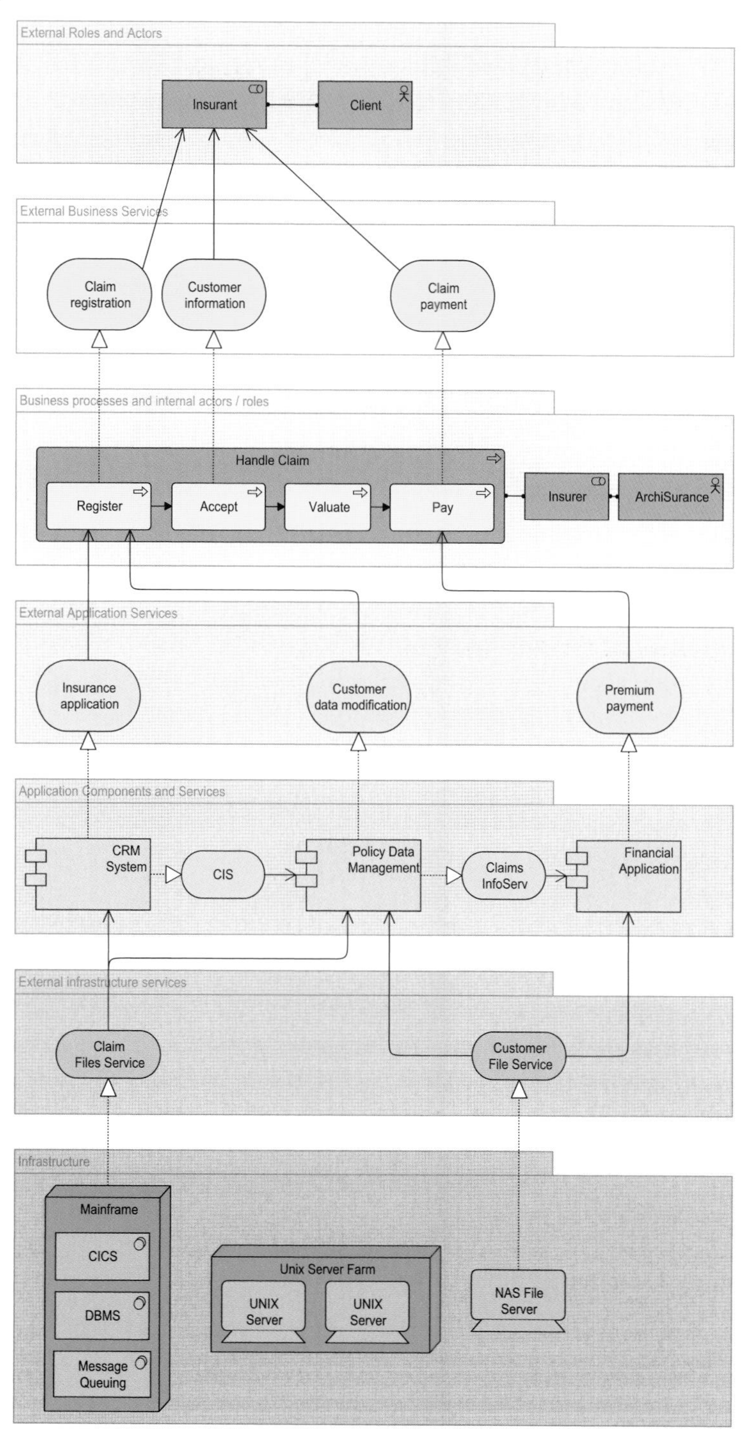
External Roles and Actors
Insurant
Client
External Business Services
Claim registration
Customer information
Claim payment
Business processes and internal actors / roles
Handle Claim
Register
Accept
Valuate
Pay
Insurer
ArchiSurance
External Application Services
Insurance application
Customer data modification
Premium payment
Application Components and Services
CRM System
CIS
Policy Data Management
Claims InfoServ
Financial Application
External infrastructure services
Claim Files Service
Customer File Service
Infrastructure
Mainframe
CICS
DBMS
Message Queuing
Unix Server Farm
UNIX Server
UNIX Server
NAS File Server

8.4.18 Landscape Map Viewpoint

A landscape map is a matrix that represents a three-dimensional co-ordinate system that represents architectural relationships. The dimensions of the landscape maps can be freely chosen from the architecture that is being modeled. In practice, often dimensions are chosen from different architectural domains; for instance, business functions, application components, and products. Note that a landscape map uses the ArchiMate *concepts*, but not the standard *notation* of these concepts.

In most cases, the vertical axis represents behavior like business processes or functions; the horizontal axis represents "cases" for which those functions or processes must be executed, such as different products, services market segments, or scenarios; the third dimension represented by the cells of the matrix is used for assigning resources like information systems, infrastructure, or human resources. The value of cells can be visualized by means of colored rectangles with text labels. Obviously, landscape maps are a more powerful and expressive representation of relationships than traditional cross tables. They provide a practical manner for the generation and publication of overview tables for managers, process, and system owners. Furthermore, architects may use landscape maps as a resource allocation instrument and as an analysis tool for the detection of patterns and changes in this allocation.

Table 24: Landscape Map Viewpoint Description

Landscape Map Viewpoint	
Stakeholders	Enterprise architects, top managers: CEO, CIO
Concerns	Readability, management and reduction of complexity, comparison of alternatives
Purpose	Deciding
Abstraction Level	Overview
Layer	Business layer, application layer, technology layer (see also Figure 4)
Aspects	Information, behavior, structure (see also Figure 4)

Concepts and Relationships

All concepts and relationships.

Example

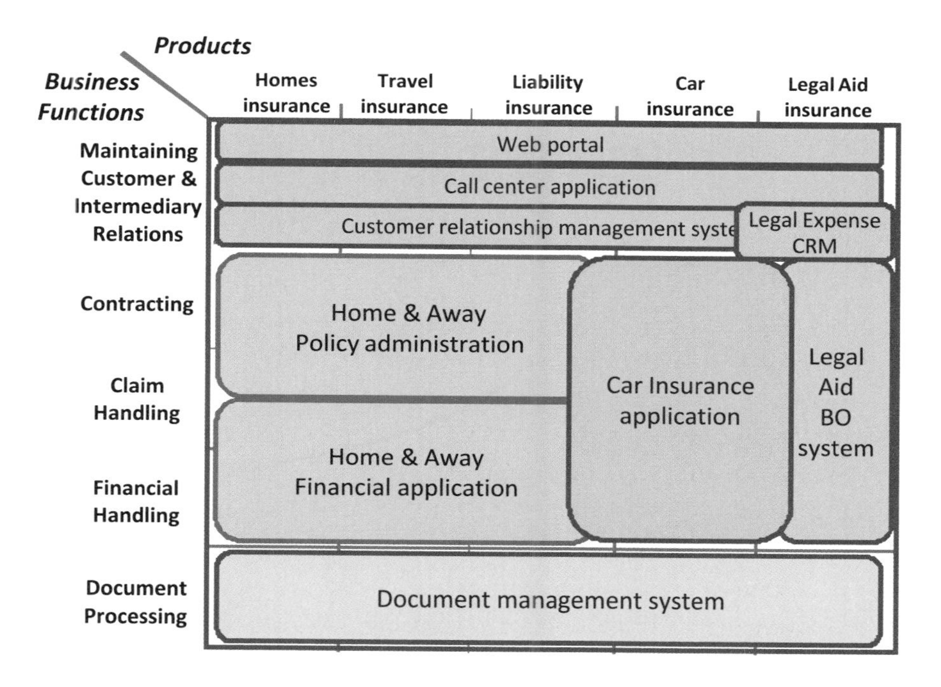

Chapter 9

Language Extension Mechanisms

Every specific purpose and usage of an architecture modeling language brings about its own specific demands on the language. Yet, it should be possible to use a language for only a limited, though non-specific, modeling purpose. Therefore, the ArchiMate core language, embedded in the ArchiMate metamodel, as described in Chapters 2 to 7, contains only the basic concepts and relationships that serve general enterprise architecture modeling purposes. However, the language should also be able to facilitate, through extension mechanisms, specialized, or domain-specific purposes, such as:

- Support for specific types of model analysis
- Support the communication of architectures
- Capture the specifics of a certain application domain (e.g., the financial sector)

The argument behind this statement is to provide a means to allow extensions of the core language that are tailored towards such specific domains or applications, without burdening the core with a lot of additional concepts and notation which most people would barely use. The remainder of this section is devoted to a number of possible extensions mechanisms that, in addition to the core, are or can become part of the ArchiMate language.

9.1 Adding Attributes to ArchiMate Concepts and Relationships

As said before, the core of ArchiMate contains only the concepts and relationships that are necessary for general architecture modeling. However, users might want to be able to, for example, perform model-based performance or cost calculations, or to attach supplementary information (textual, numerical, etc.) to the model elements. A simple way to enrich ArchiMate concepts and relationships in a generic way is to add supplementary information by means of a "profiling" specialization mechanism (see also [11]). A *profile* is a data structure which can be defined separate from the ArchiMate language, but can be dynamically coupled with concepts or relationships; i.e., the user of the language is free to decide whether and when the assignment of a profile to a model element is necessary. Profiles can be specified as sets of typed attributes, by means of a profile definition language. Each of these attributes may have a default value that can be changed by the user.

We can distinguish two types of profiles:

- *Pre-defined profiles*: These are profiles that have a predefined attribute structure and can be implemented beforehand in any tool supporting the ArchiMate language. Examples of such profiles are sets of attributes for ArchiMate concepts and relationships that have to be specified in order to execute common types of analysis.
- *User-defined profiles:* Through a profile definition language, the user is able to define new profiles, thus extending the definition of ArchiMate concepts or relationships with supplementary attribute sets.

Example

Table 25 below shows possible profiles with input attributes needed for certain types of cost and performance analysis of architecture models [19]. Each "used by" relationship may have a weight (indicating the average number of uses); each (business, application, or infrastructure) "service" may have fixed and variable costs and an (average) service time; and each structure element (e.g., business role, business actor, application component, device) may have fixed and variable costs and a capacity.

Table 25: Profile Example

"Used By" Profile		**"Service" Profile**		**"Structure Element" Profile**	
Attribute	**Type**	**Attribute**	**Type**	**Attribute**	**Type**
Weight	Real	Fixed cost	Currency	Fixed cost	Currency
		Variable cost	Currency	Variable cost	Currency
		Service time	Time	Capacity	Integer

9.2 Specialization of Concepts

Specialization is a simple and powerful way to define new concepts based on the existing ones. Specialized concepts inherit the properties of their "parent" concepts, but additional restrictions with respect to their use may apply. For example, some of the relationships that apply for the "parent" concept need not be allowed for the specialization. A specialized concept strongly resembles a stereotype as it is used in UML. Specialization of concepts provides extra flexibility, as it allows organizations or individual users to customize the language to their own preferences and needs, while the underlying precise definition of the concepts is conserved. This also implies that analysis and visualization techniques developed for the ArchiMate language still apply when the specialized concepts are used.

Figure 60 shows a number of examples of concept specializations that have proven to be useful in several practical cases.

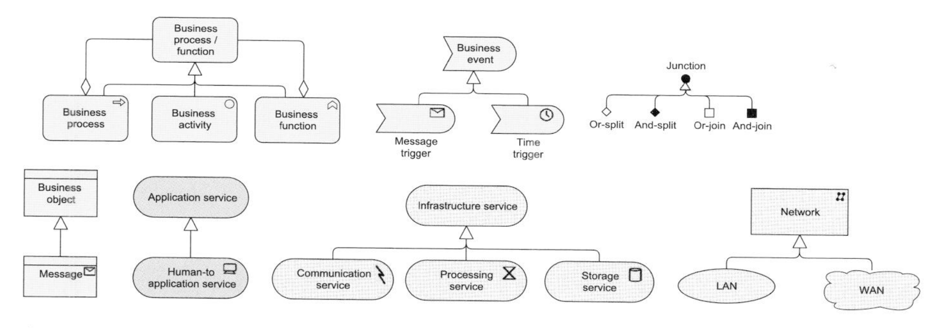

Figure 60: More Examples of Specialized Concepts

Also, the concepts in the layer-specific metamodels can be considered specializations of the concepts in the generic metamodel of Chapter 2.

As the above examples indicate, we may introduce a new graphical notation for a specialized concept, but usually with a resemblance to the notation of the parent concept; e.g., by adding or changing the icon. It is also possible to use a <<stereotype>>-notation as in UML. Finally, for a specialized concept, certain attributes may be predefined, as described in the previous section.

Chapter 10

Motivation Extension

10.1 Motivation Aspect Metamodel

Figure 61 shows the metamodel of motivational concepts. It includes the actual motivations or intentions – i.e., goals, principles, requirements, and constraints – and the sources of these intentions; i.e., stakeholders, drivers, and assessments.

Motivational elements are related to the core elements via the requirement or constraint concept.

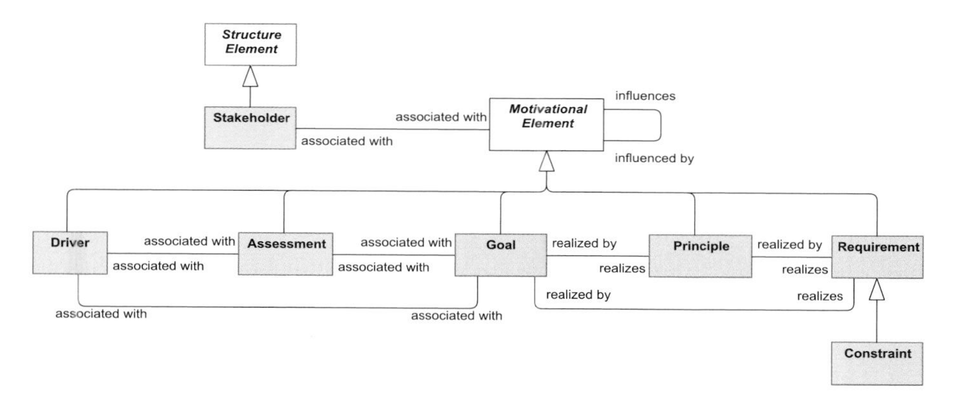

Figure 61: Motivation Extension Metamodel

Note: This figure does not show all permitted relationships: every non-abstract element in the Motivation extension can have aggregation and specialization relationships with elements of the same type.

10.2 Motivational Concepts

Motivational concepts are used to model the motivations, or reasons, that underlie the design or change of some enterprise architecture. These motivations influence, guide, and constrain the design.

It is essential to understand the factors, often referred to as *drivers*, which influence the motivational elements. They can originate from either inside or outside the enterprise. Internal drivers, also called *concerns*, are associated with *stakeholders*, which can be some individual human being or some group of human beings, such as a project team, enterprise, or society. Examples of such internal drivers are customer satisfaction, compliance to legislation, or profitability. It is common for enterprises to undertake an *assessment* of these drivers; e.g., using a SWOT analysis, in order to respond in the best way.

The actual motivations are represented by goals, principles, requirements, and constraints. *Goals* represent some desired result – or end – that a stakeholder wants to achieve; e.g., increasing customer satisfaction by 10%. Principles and requirements represent desired properties of solutions – or means – to realize the goals. *Principles* are normative guidelines that guide the design of all possible solutions in a given context. For example, the principle "Data should be stored only once" represents a means to achieve the goal of "Data consistency" and applies to all possible designs of the organization's architecture. *Requirements* represent formal statements of need, expressed by stakeholders, which must be met by the architecture or solutions. For example, the requirement "Use a single CRM system" conforms to the aforementioned principle by applying it to the current organization's architecture in the context of the management of customer data.

10.2.1 Stakeholder

A stakeholder is defined as the role of an individual, team, or organization (or classes thereof) that represents their interests in, or concerns relative to, the outcome of the architecture.

This definition is based on the definition in TOGAF [4]. A stakeholder has one or more interests in, or concerns about, the organization and its enterprise architecture. In order to direct efforts to these interests and concerns, stakeholders change, set, and emphasize goals. Examples of stakeholders are the CEO, the board of directors, shareholders, customers, business, and application architects, but also legislative authorities. The name of a stakeholder should preferably be a noun.

Figure 62: Stakeholder Notation

Example

The model below illustrates the modeling of stakeholders. Two main stakeholders are modeled: the Board of ArchiSurance and Customer. The Board is composed of three other stakeholders: the CIO, the CEO, and the CFO.

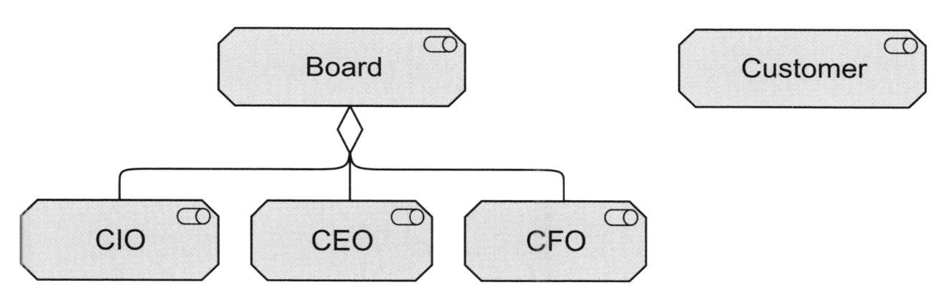

Example 47: Stakeholder

10.2.2 Driver

A driver is defined as something that creates, motivates, and fuels the change in an organization.

Drivers may be internal, in which case they are usually associated with a stakeholder. Examples of internal drivers (or "concerns") are "Customer satisfaction", "Compliance to legislation", and "Profitability". Drivers of change may also be external; e.g., economic changes or changing legislation. The name of a driver should preferably be a noun.

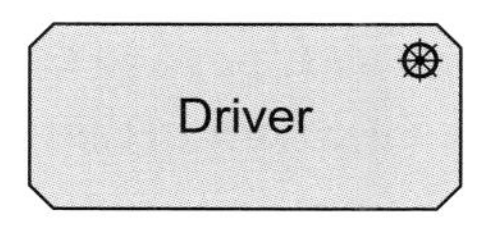

Figure 63: Driver Notation

Example

The model below illustrates the modeling of internal and external drivers of change. Stakeholders CEO and Customer share a common concern Customer satisfaction, which is an internal driver of change. The stakeholder CEO also has the satisfaction of the company's shareholders as a concern. This driver can be decomposed into two sub-drivers: Profit and Stock value. In addition to these internal drivers, there is an external driver Economic changes, which influences the stock value.

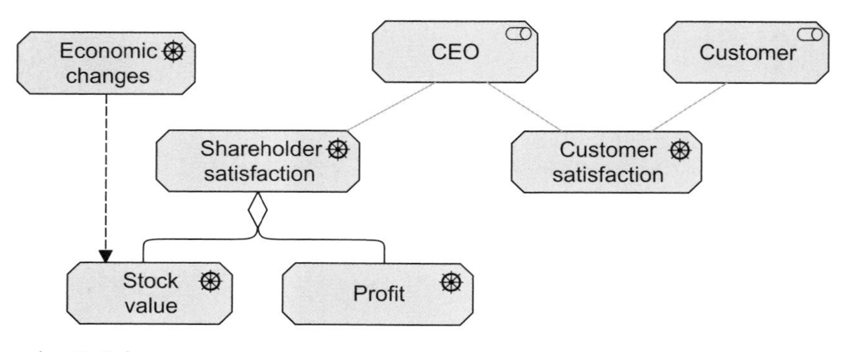

Example 48: Driver

10.2.3 Assessment

An assessment is defined as the outcome of some analysis of some driver.

An assessment may reveal strengths, weaknesses, opportunities, or threats for some area of interest. These outcomes need to be addressed by adjusting existing goals or setting new ones, which may trigger changes to the enterprise architecture.

Strengths and weaknesses are internal to the organization. Opportunities and threats are external to the organization. Weaknesses and threats can be considered as problems that need to be addressed by goals that "negate" the weaknesses and threats. Strengths and opportunities may be translated directly into goals. For example, the weakness "customers complain about the helpdesk" can be addressed by defining the goal "improve helpdesk". Or, the opportunity "customers favor insurances that can be managed on-line" can be addressed by the goal "introduce on-line portfolio management". The name of an assessment should preferably be a noun or a (very) short sentence.

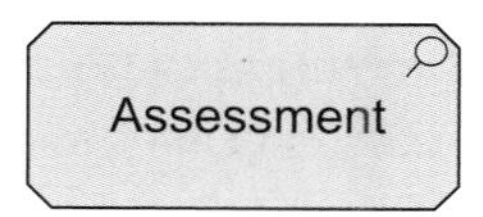

Figure 64: Assessment Notation

Example

The model below describes the assessments of driver Customer satisfaction and the sub-concern Helpdesk support. In this case, all assessments represent weaknesses. Concerning Customer satisfaction in general, customers complain and even leave ArchiSurance. The assessment Complaining customers is further detailed and divided into four complaints: the lack of insight into the status of a claim, the inconvenient way of submitting claims, the lack of insight into the customer's portfolio, and the inconsistency and incompleteness of customer information. Concerning Helpdesk support in particular, customers experience long waiting queues and high service times.

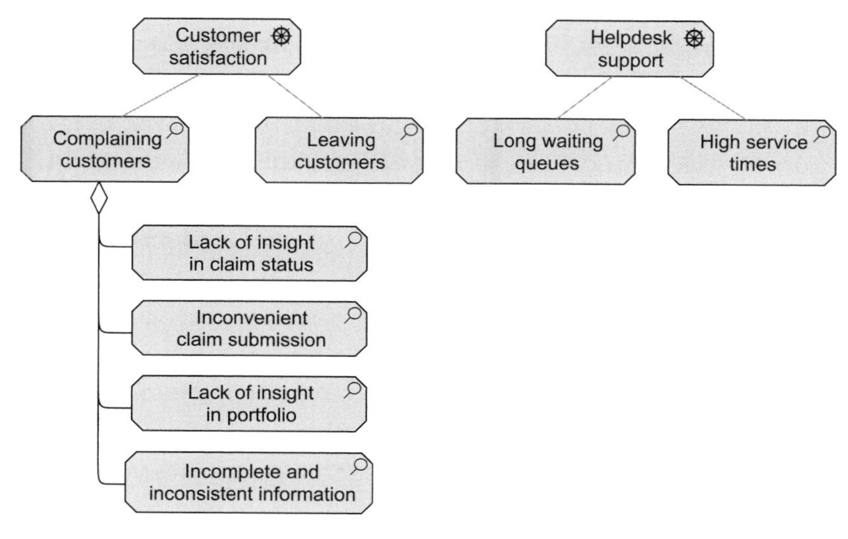

Example 49: Assessment

10.2.4 Goal

A goal is defined as an end state that a stakeholder intends to achieve.

In principle, an end can represent anything a stakeholder may desire, such as a state of affairs, or a produced value. Examples of goals are: to increase profit, to reduce waiting times at the helpdesk, or to introduce on-line portfolio management.

Goals are generally expressed using qualitative words; e.g., "increase", "improve", or "easier". Goals can also be decomposed; e.g., "increase profit" can be decomposed into the goals "reduce cost" and "increase sales". However, it is also very common to associate concrete objectives with goals, which can be used to describe both the quantitative and time-related measures which are essential to describe the desired state, and when it should be achieved.

Figure 65: Goal Notation

Example
The model below illustrates the modeling of goals to address the assessments of the driver Costs: the applications costs and the costs of employees are too high. The former assessment is addressed by the goals Reduce maintenance costs and Reduce direct application costs (of usage). The latter assessment is addressed by the goal Reduce workload employees, which is decomposed into Reduce manual work and Reduce interaction with customer.

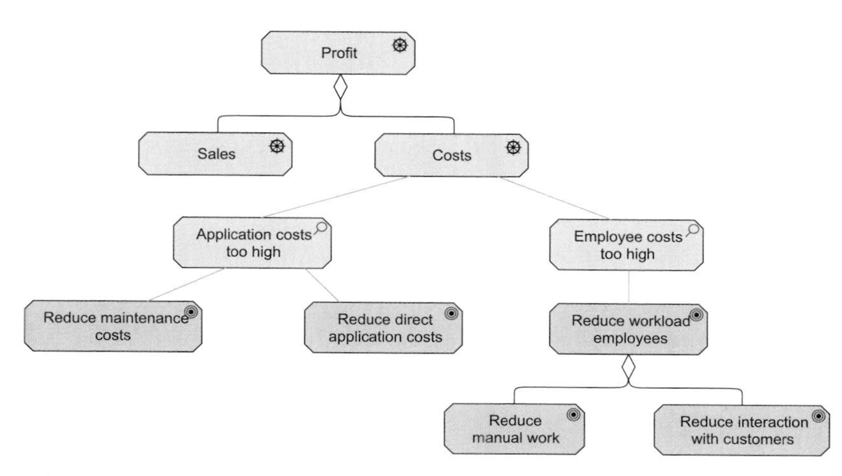

Example 50: Goal

10.2.5 Requirement

> A requirement is defined as a statement of need that must be realized by a system.

In the end, a business goal must be realized by a plan or concrete change goal, which may or may not require a new system or changes to an existing system.

The term "system" is used in its general meaning; i.e., as a group of (functionally) related elements, where each element may be considered as a system again. Therefore, a system may refer to any active structural element, behavioral element, or passive structural element of some organization, such as a business actor, application component, business process, application service, business object, or data object.

Requirements model the properties of these elements that are needed to achieve the "ends" that are modeled by the goals. In this respect, requirements represent the "means" to realize goals.

During the design process, goals may be decomposed until the resulting sub-goals are sufficiently detailed to enable their realization by properties that can be exhibited by systems. At this point, goals can be realized by requirements that assign these properties to the systems.

For example, one may identify two alternative requirements to realize the goal to improve portfolio management: (i) by assigning a personal assistant to each customer, or (ii) by introducing on-line portfolio management. The former requirement can be realized by a human actor and the latter by a software application. These requirements can be decomposed further to define the requirements on the human actor and the software application in more detail.

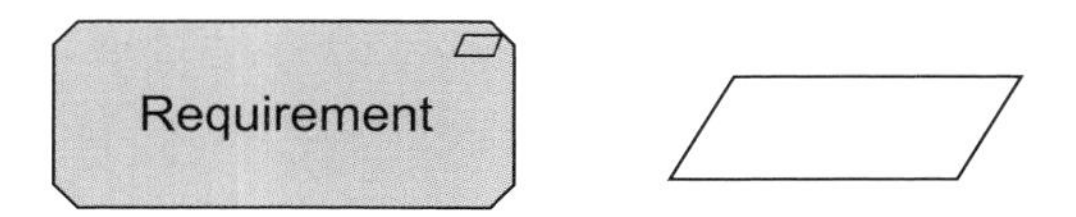

Figure 66: Requirement Notation

Example

The model below illustrates the decomposition of goals towards requirements. The goals Facilitate self-service and Make customer interaction more effective result from the successive decomposition of the goals Reduce workload employees and Reduce interaction with customer. The goal Facilitate self-service can be realized by the alternative requirements Provide on-line portfolio service and Provide on-line information service. Both requirements are realized by some software application. In addition, the requirement Provide on-line portfolio service may realize the goal Improve portfolio management. Alternatively, this goal can be realized by assigning a personal assistant to each customer.

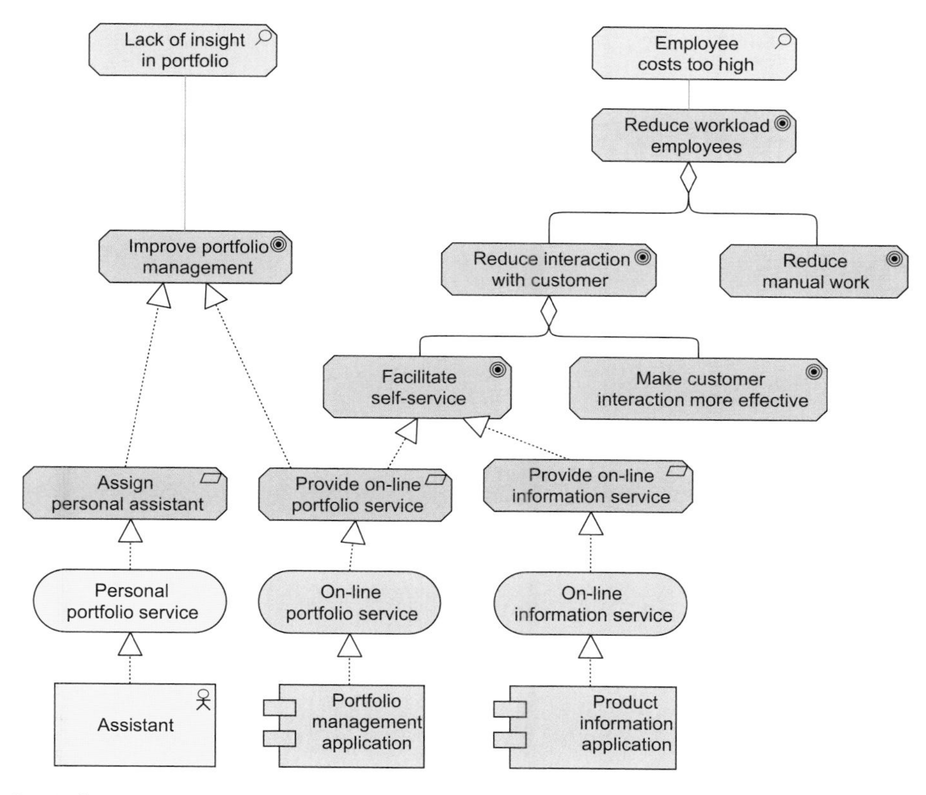

Example 51: Requirement

10.2.6 Constraint

A constraint is defined as a restriction on the way in which a system is realized.

In contrast to a requirement, a constraint does not prescribe some intended functionality of the system to be realized, but imposes a restriction on the way in which the system may be realized. This may be a restriction on the implementation of the system (e.g., specific technology that is to be used), or a restriction on the implementation process (e.g., time or budget constraints).

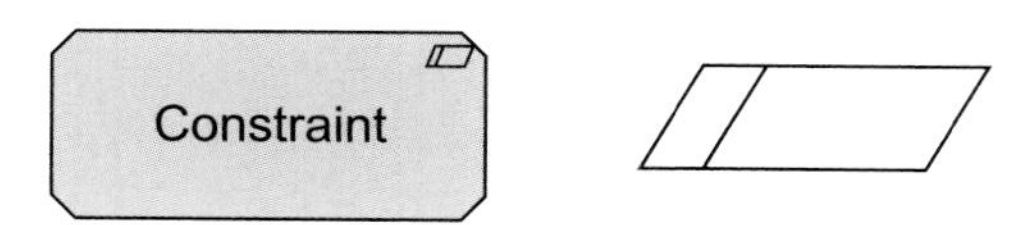

Figure 67: Constraint Notation

Example

For the realization of a new portfolio management application, two constraints are imposed, as shown in the model below: for the realization of the application, Java should be used, and the budget of the implementation project is limited to 500k Euro.

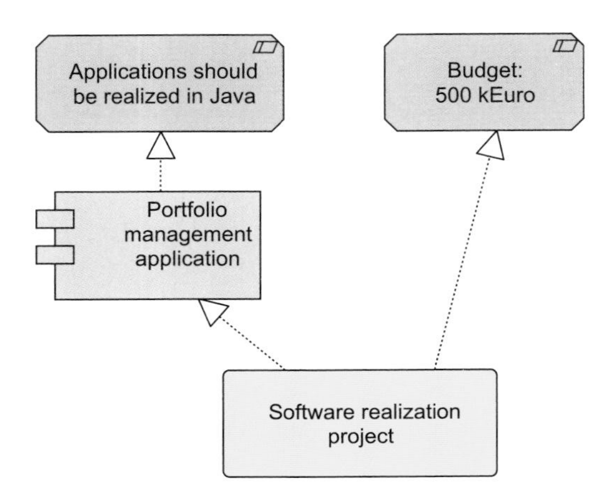

Example 52: Constraint

10.2.7 Principle

A principle is defined as a normative property of all systems in a given context, or the way in which they are realized.

Principles are strongly related to goals and requirements. Similar to requirements, principles define intended properties of systems. However, in contrast to requirements, principles are broader in scope and more abstract than requirements. A principle defines a general property that applies to any system in a certain context. A requirement defines a property that applies to a specific system.

A principle needs to be made specific for a given system by means of one or more requirements, in order to enforce that the system conforms to the principle. For example, the principle "Information management processes comply with all relevant laws, policies, and regulations" is realized by the requirements that are imposed by the actual laws, policies, and regulations that apply to the specific system under design.

A principle is motivated by some goal. For example, the aforementioned principle may be motivated by the goal to maintain a good reputation and/ or the goal to avoid penalties. The principle provides a means to realize its motivating goal, which is generally formulated as a guideline. This guideline constrains the design of all systems in a given context by stating the general properties that are required from any system in this context to realize the goal. Principles are intended to be more stable than requirements in the sense that they do not change as quickly as requirements may do. Organizational values, best practices, and design knowledge may be reflected and made applicable in terms of principles.

Figure 68: Principle Notation

Example
The model below illustrates the use of principles. Principle Systems should be customer facing is modeled as a means to realize the goals Reduce interaction with customer and Reduce manual work. The principle is further specialized into the requirements Provide on-line portfolio service and Provide on-line information service to apply the principle to the actual systems (architecture) under design.

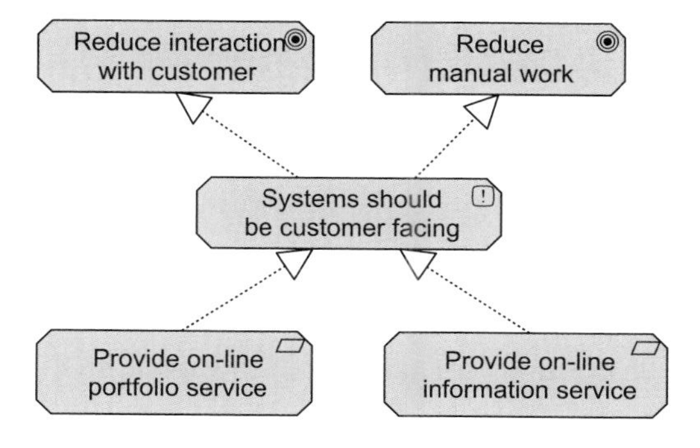

Example 53: Principle

10.2.8 Summary of Motivational Concepts

gives an overview of the motivational concepts, with their definitions.

Table 26: Motivational Concepts

Concept	Definition	Notation
Stakeholder	The role of an individual, team, or organization (or classes thereof) that represents their interests in, or concerns relative to, the outcome of the architecture.	Stakeholder
Driver	Something that creates, motivates, and fuels the change in an organization.	Driver
Assessment	The outcome of some analysis of some driver.	Assessment
Goal	An end state that a stakeholder intends to achieve.	Goal
Requirement	A statement of need that must be realized by a system.	Requirement
Constraint	A restriction on the way in which a system is realized.	Constraint
Principle	A normative property of all systems in a given context, or the way in which they are realized.	Principle

10.3 Relationships

The metamodels and examples from the previous sections show the different types of relationships that can be used between two motivational elements and between one motivational element and one core element. This section provides a more precise description of these relationships.

10.3.1 Aggregation Relationship

The aggregation relationship models that some intention is divided into multiple intentions.

The aggregation relationship is generally used to describe an intention in more detail by decomposing the intention into multiple, more concrete intentions.

Figure 69: Aggregation Notation

Alternatively, an aggregation can be expressed by nesting the model elements.

Example
The models below show the two ways to express the decomposition of goal **Reduce workload employees** into the sub goals Reduce interaction with customer and Reduce manual work.

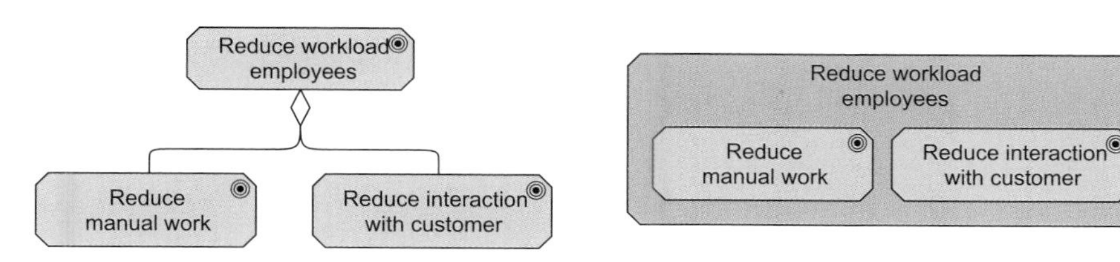

Example 54: Aggregation (Decomposition)

10.3.2 Realization Relationship

> The realization relationship models that some end is realized by some means.

The realization relationship is used to represent the following means-end relationships:

1. A goal (the end) is realized by a principle, constraint, or requirement (the means).
2. A principle (the end) is realized by a constraint or requirement (the means).
3. A requirement (the end) is realized by a system (the means), which can be represented by an active structure element, a behavior element, or a passive structure element.

Figure 70: Realization Notation

> **Example**
>
> The model below illustrates several ways to use the realization relationship. Principle Systems should be customer facing is a means to realize the goal Reduce interaction with customer. Requirement Provide on-line portfolio service is a means to realize sub-goal Facilitate self-service, and to realize the principle Systems should be customer facing. And this requirement can be realized by the business service On-line portfolio service.

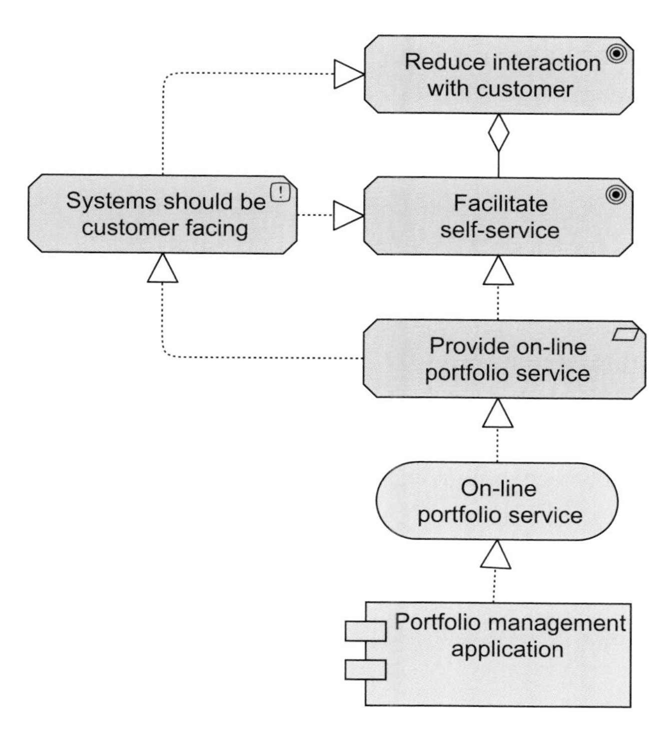

Example 55: Realization (Means-End)

10.3.3 Influence Relationship

> The influence relationship models that some motivational element has a positive or negative influence on another motivational element.

The influence relationship is used to describe that some motivational element may influence (the realization of) another motivational element. In general, a motivational element is realized to a certain degree. An influence by some other motivational element may affect this degree positively or negatively, depending on the degree in which this other motivational element is satisfied itself. For example, the degree in which the goal to increase customer satisfaction is realized may be represented by the percentage of satisfied customers that participate in a market interview. This percentage may be influenced positively by, for example, the goal to improve the company's reputation; i.e., a higher degree of improvement results in a higher increase in customer satisfaction. On the other hand, the goal to lay off employees may influence the company's reputation negatively; i.e., more lay-offs could result in a lower increase (or even decrease) in the company's reputation. And thus (indirectly), the goal to increase customer satisfaction may also be influenced negatively.

A positive influence relationship does not imply that the realization of the influenced motivational element depends on the contributing intention. The necessary means to realize some motivational element are modeled using the realization relationship.

A negative influence relationship does not imply that the realization of the influenced motivational element is completely excluded by the contributing motivational element.

The influence relationship re-uses the notation of the flow relationship, signifying a "flow of influence". An attribute can be used to indicate the direction and strength of the influence. The choice of possible attribute values is left to the modeler; e.g., {++, +, 0, -, --} or [0..10].[1]

Figure 71: Influence Notation

Example

The model below illustrates the use of the influence relationship for making a trade-off between the two requirements that realize the goal Improve portfolio management. The goal Increase customer satisfaction and the principle Systems should be customer facing are used as trade-off criteria. Both requirements positively influence the intended increase of customer satisfaction. The requirement of using a personal assistant scores a little better for this criterion. However, the requirement scores a lot worse for the customer-facing criterion. The positive score of the requirement Provide on-line portfolio service for the customer-facing principle is consistent with the description of the requirement realizing the principle in an earlier example.

1 This standard abstracts from the specification of the functions that describe the exact relation between the degree of realization of the related intentions and the strength factor.

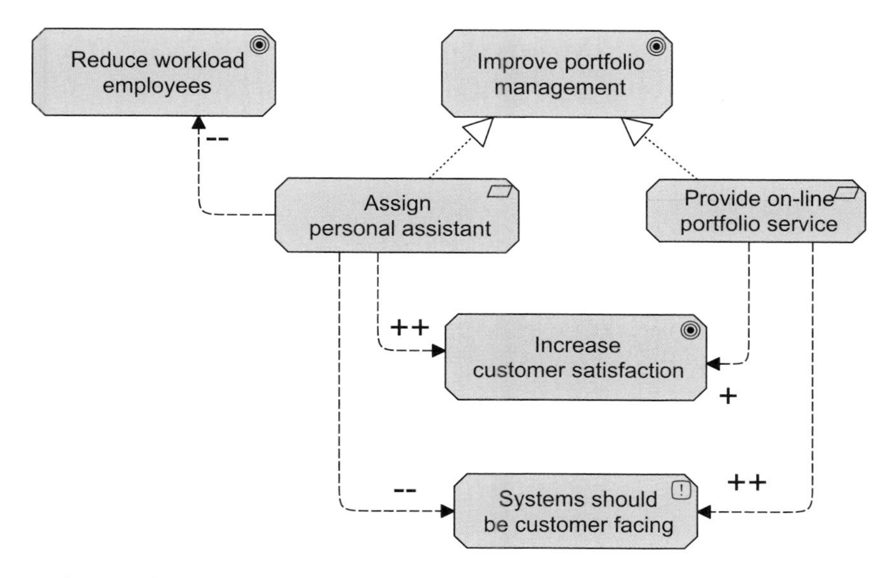

Example 56: Influence

10.3.4 Summary of Relationships

Table 27 gives an overview of the relationships, with their definition, that involve one or more intentional concepts.

Table 27: Relationships

Intentional Relationships		**Notation**
Aggregation	Aggregation models that some intentional element is divided into multiple intentional elements.	
Realization	Realization models that some end is realized by some means.	
Influence	Influence models that some motivational element has a positive or negative influence on the realization of another motivational element.	

10.4 Cross-Aspect Dependencies

The purpose of the motivation extension is to model the motivation behind the core elements in some enterprise architecture. Therefore, it should be possible to relate motivational elements to core elements.

As shown in Figure 72, a requirement or constraint can be related directly to a core element by means of a realization relationship. Other motivational elements cannot be related directly to core elements, but only indirectly by means of derived relationships via requirements or constraints.

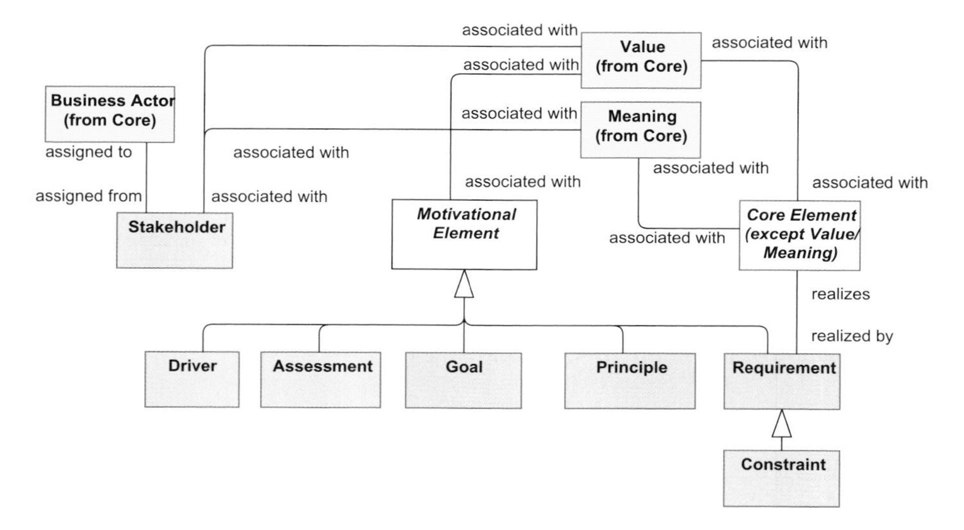

Figure 72: Relationships between Motivation Extension and the ArchiMate Core Concepts

Also, a business actor may be assigned to a stakeholder, which can be seen as a motivational role (as opposed to an operational business role) that an actor may fulfill.

10.5 Viewpoints

A number of standard viewpoints for modeling motivational aspects have been defined. Each of these viewpoints presents a different perspective on modeling the motivation that underlies some enterprise architecture and allows a modeler to focus on certain aspects. Therefore, each viewpoint considers only a selection of the concepts and relationships that have been described in the preceding sections.

The following viewpoints are distinguished:

- The *stakeholder viewpoint*, which focuses on modeling the stakeholders, drivers, the assessments of these drivers, and the initial goals to address these drivers and assessments
- The *goal realization viewpoint*, which focuses on refining the initial, high-level goals into more concrete (sub-)goals using the aggregation relationship, and finally into requirements and constraints using the realization relationship
- The *goal contribution viewpoint*, which focuses on modeling and analyzing the influence relationships between goals (and requirements)
- The *principles viewpoint*, which focuses on modeling the relevant principles and the goals that motivate these principles
- The *requirements realization viewpoint*, which focuses on modeling the realization of requirements and constraints by means of core elements, such as actors, services, processes, application components, etc.
- The *motivation viewpoint*, which covers the entire motivational aspect and allows one to use all motivational elements

All viewpoints are separately described below. For each viewpoint the comprised concepts and relationships, the guidelines for the viewpoint use, and the goal and target group and of the viewpoint are indicated. Furthermore, each viewpoint description contains example models. For more details on the goal and use of viewpoints, refer to [2], Chapter 8.

10.5.1 Stakeholder Viewpoint

The stakeholder viewpoint allows the analyst to model the stakeholders, the internal and external drivers for change, and the assessments (in terms of strengths, weaknesses, opportunities, and threats) of these drivers. Also, the links to the initial (high-level) goals that address these concerns and assessments may be described. These goals form the basis for the requirements engineering process, including goal refinement, contribution and conflict analysis, and the derivation of requirements that realize the goals.

Table 28: Stakeholder Viewpoint Description

Stakeholder Viewpoint	
Stakeholders	Stakeholders, business managers, enterprise and ICT architects, business analysts, requirements managers
Concerns	Architecture mission and strategy, motivation
Purpose	Designing, deciding, informing
Abstraction Level	Coherence, Details
Layer	Business, Application, and Technology layers
Aspects	Motivation

Concepts and Relationships

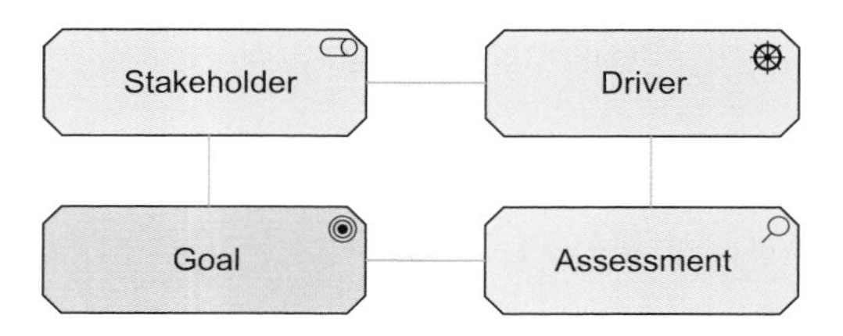

Example

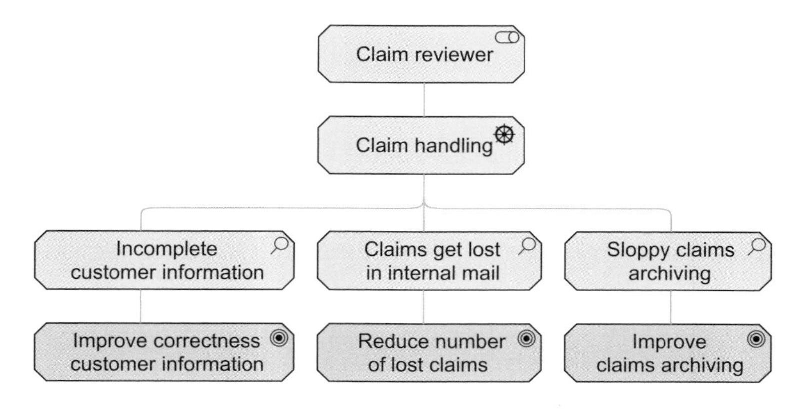

10.5.2 Goal Realization Viewpoint

The goal realization viewpoint allows a designer to model the refinement of (high-level) goals into more concrete goals, and the refinement of concrete goals into requirements or constraints that describe the properties that are needed to realize the goals. The refinement of goals into sub-goals is modeled using the aggregation relationship. The refinement of goals into requirements is modeled using the realization relationship.

In addition, the principles may be modeled that guide the refinement of goals into requirements.

Table 29: Goal Realization Viewpoint Description

Goal Realization Viewpoint	
Stakeholders	Stakeholders, business managers, enterprise and ICT architects, business analysts, requirements managers
Concerns	Architecture mission, strategy and tactics, motivation
Purpose	Designing, deciding
Abstraction Level	Coherence, Details
Layer	Business, Application, and Technology layers
Aspects	Motivation

Concepts and Relationships

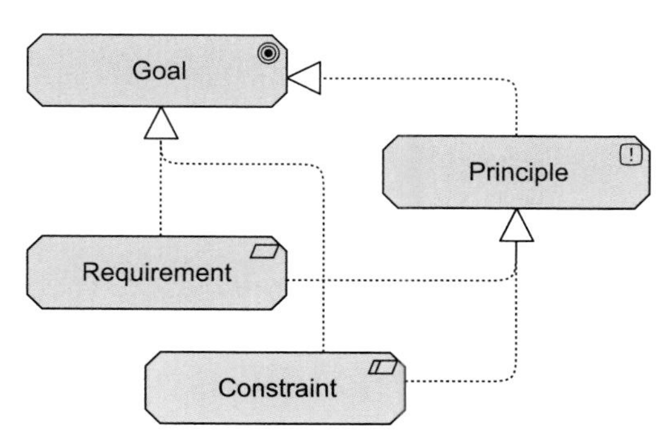

Example

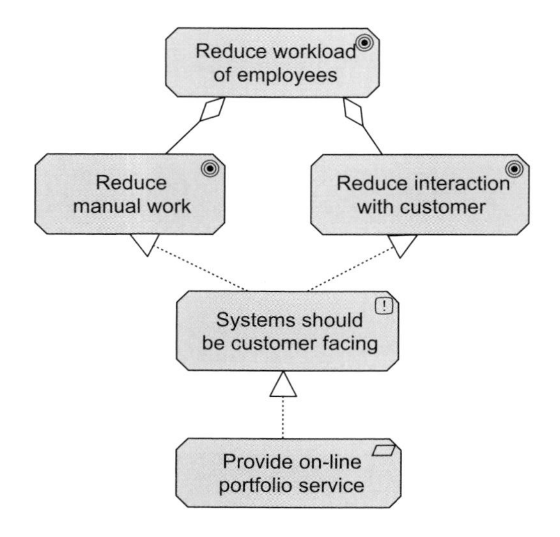

10.5.3 Goal Contribution Viewpoint

The goal contribution viewpoint allows a designer or analyst to model the influence relationships between goals and requirements. The resulting views can be used to analyze the impact that goals have on each other or to detect conflicts between stakeholder goals.

Typically, this viewpoint may be used after goals have, to some extent, been refined into sub-goals and, possibly, into requirements. Therefore, aggregation and realization relationships may also be shown in this viewpoint.

Table 30: Goal Contribution Description

Goal Contribution Viewpoint	
Stakeholders	Stakeholders, business managers, enterprise and ICT architects, business analysts, requirements managers
Concerns	Architecture mission, strategy and tactics, motivation

Goal Contribution Viewpoint		
Purpose	Designing, deciding	
Abstraction Level	Coherence, Details	
Layer	Business, Application, and Technology layers	
Aspects	Motivation	

Concepts and Relationships

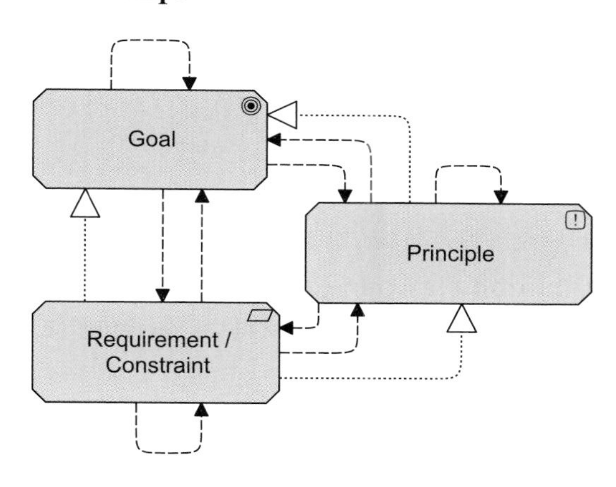

Example

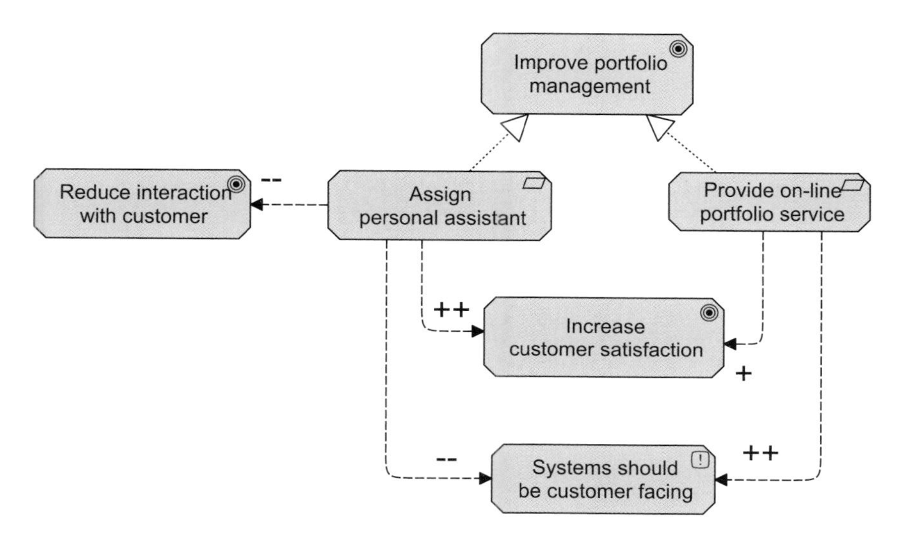

10.5.4 Principles Viewpoint

The principles viewpoint allows the analyst or designer to model the principles that are relevant to the design problem at hand, including the goals that motivate these principles. In addition, relationships between principles, and their goals, can be modeled. For example, principles may influence each other positively or negatively.

Table 31: Principles Viewpoint Description

Principles Viewpoint	
Stakeholders	Stakeholders, business managers, enterprise and ICT architects, business analysts, requirements managers
Concerns	Architecture mission and strategy, motivation
Purpose	Designing, deciding, informing
Abstraction Level	Coherence, Details
Layer	Business, Application, and Technology layers
Aspects	Motivation

Concepts and Relationships

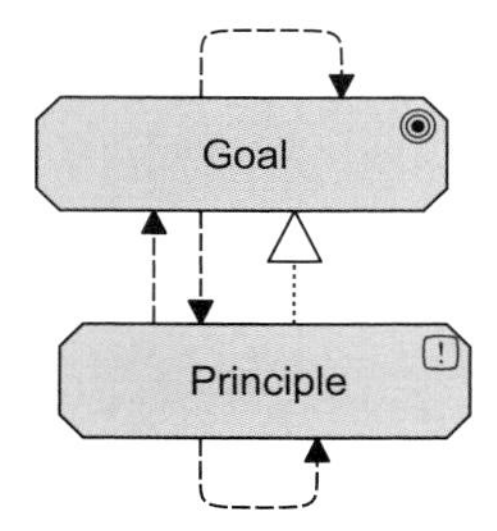

Example

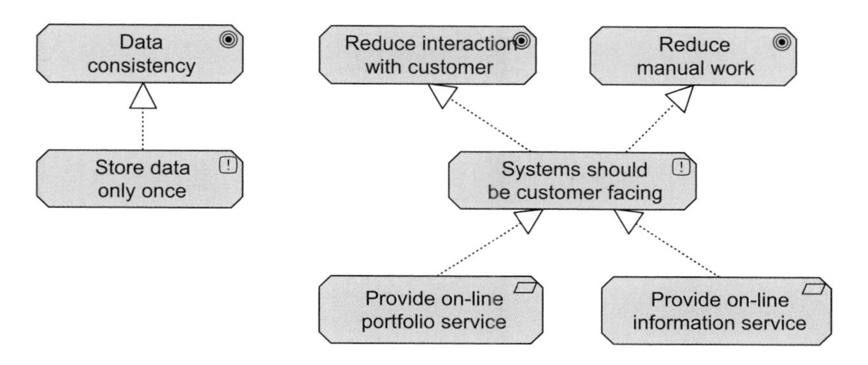

10.5.5 Requirements Realization Viewpoint

The requirements realization viewpoint allows the designer to model the realization of requirements by the core elements, such as business actors, business services, business processes, application services, application components, etc. Typically, the requirements result from the goal refinement viewpoint.

In addition, this viewpoint can be used to refine requirements into more detailed requirements. The aggregation relationship is used for this purpose.

Table 32: Requirements Realization Viewpoint Description

Requirements Realization Viewpoint	
Stakeholders	Enterprise and ICT architects, business analysts, requirements managers
Concerns	Architecture strategy and tactics, motivation
Purpose	Designing, deciding, informing
Abstraction Level	Coherence, Details
Layer	Business, Application, and Technology layers
Aspects	Motivation

Concepts and Relationships

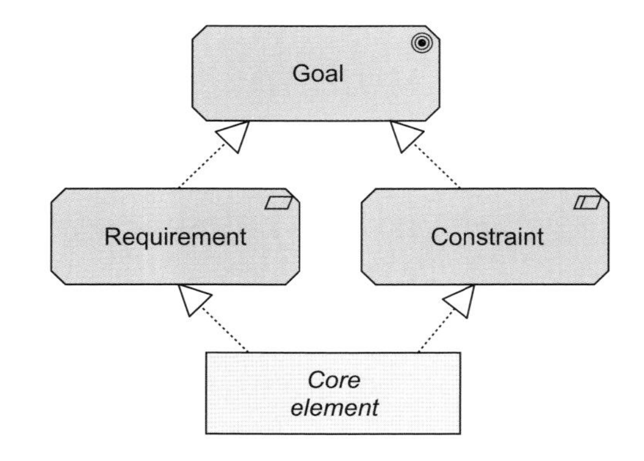

Example

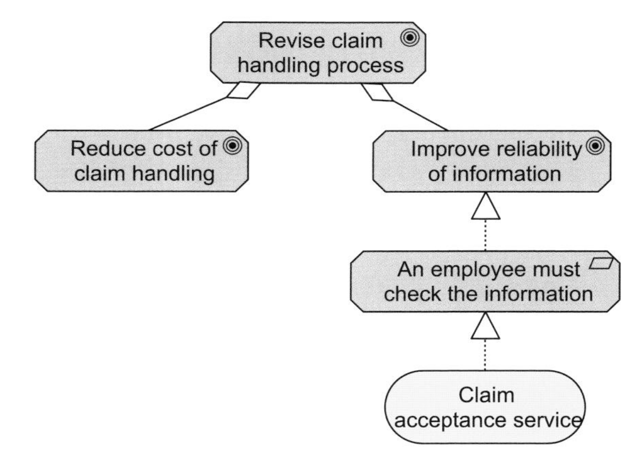

10.5.6 Motivation Viewpoint

The motivation viewpoint allows the designer or analyst to model the motivation aspect, without focusing on certain elements within this aspect. For example, this viewpoint can be used to present a complete or partial overview of the motivation aspect by relating stakeholders, their primary goals, the principles that are applied, and the main requirements on services, processes, applications, and objects.

Table 33: Motivation Viewpoint Description

Motivation Viewpoint	
Stakeholders	Enterprise and ICT architects, business analysts, requirements managers
Concerns	Architecture strategy and tactics, motivation
Purpose	Designing, deciding, informing
Abstraction Level	Overview, Coherence, Details
Layer	Business, Application, and Technology layers
Aspects	Motivation

Concepts and Relationships

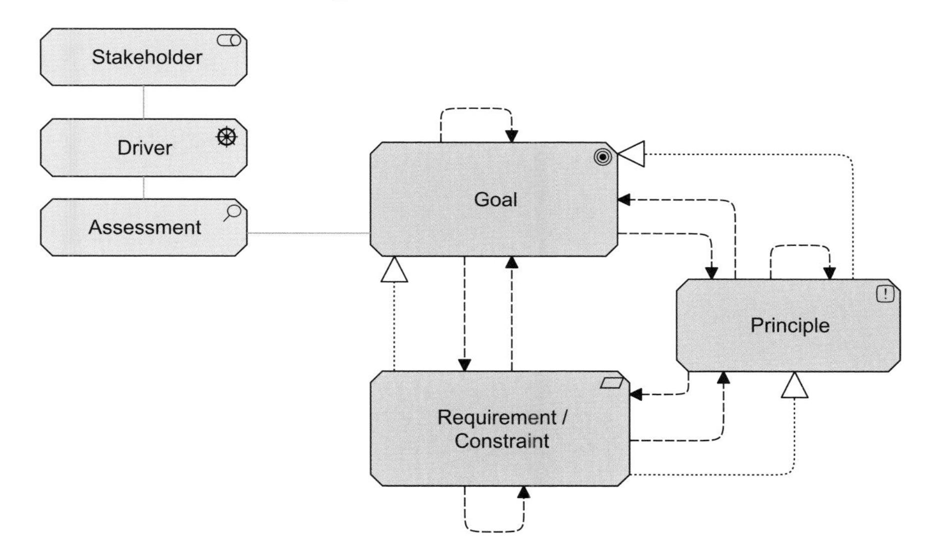

Example

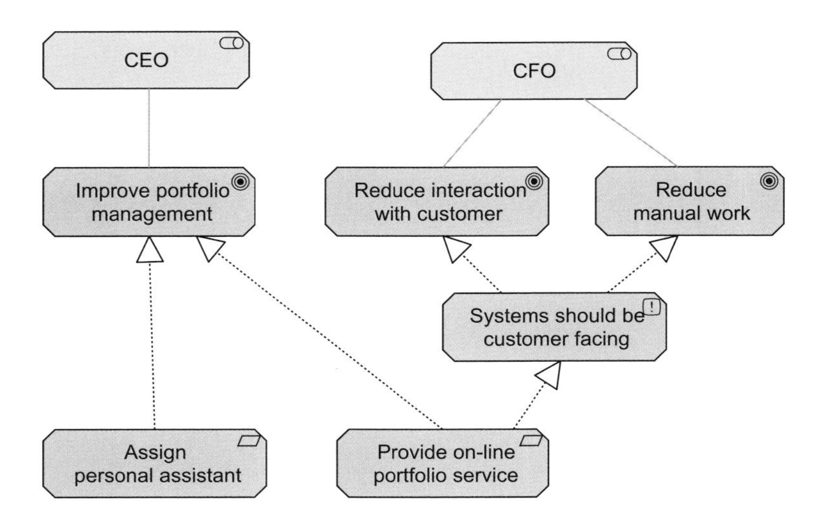
CEO
CFO
Improve portfolio management
Reduce interaction with customer
Reduce manual work
Systems should be customer facing
Assign personal assistant
Provide on-line portfolio service

Chapter 11

Implementation and Migration Extension

11.1 Implementation and Migration Extension Metamodel

Figure 73 shows the metamodel of implementation and migration concepts.

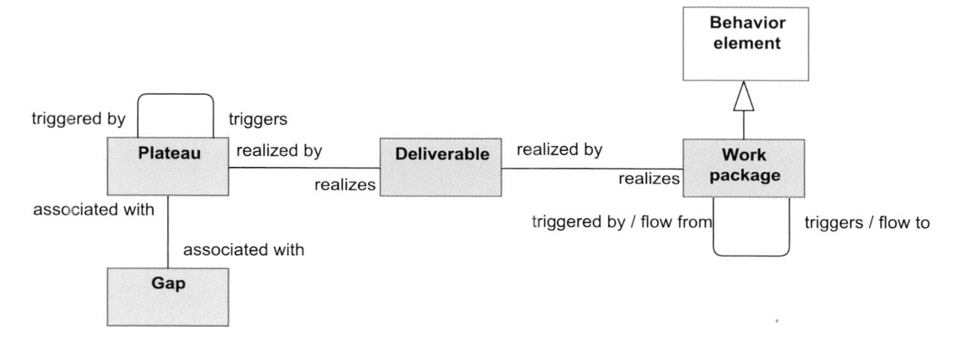

Figure 73: Implementation and Migration Extension Metamodel

Conceptually, a work package is similar to a business process, in that it consists of a set of causally related tasks, aimed at producing a well-defined result. However, a work package is a unique, "one-off" process. Still, a work package can be described in a way very similar to the description of a process.

11.2 Implementation and Migration Concepts

11.2.1 Work Package

The central behavioral concept is a *work package*. A work package has a clearly defined beginning and end date, and a well-defined set of goals or results. The work package concept can be used to model projects, but also, e.g., sub-projects or tasks within a project, programs, or project portfolios.

> A work package is defined as a series of actions designed to accomplish a unique goal within a specified time.

Work package

Figure 74: Work Package Notation

> **Example**
>
> The model below illustrates a model of a work package that models a program to rationalize the application portfolio. This program consists of two projects that are executed sequentially, each of them also modeled as a work package. First, a project is carried out to integrate the back-office systems (except for the CRM systems) into a single back-office system. Next, a project is carried out to integrate the CRM systems.

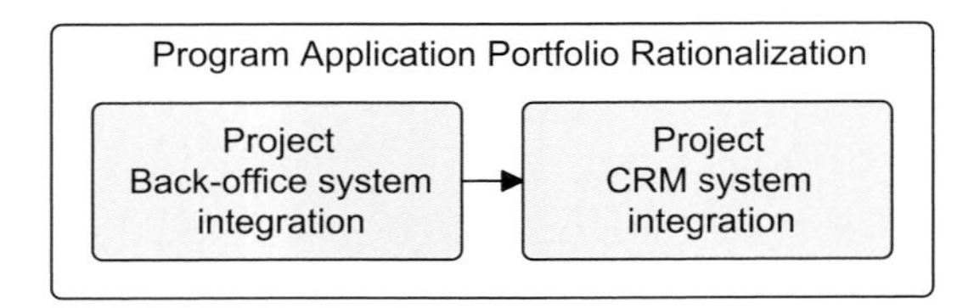

Example 57: Work Package

11.2.2 Deliverable

Work packages produce *deliverables*. These may be results of any kind; e.g., reports, papers, services, software, physical products, etc., or intangible results such as organizational change. A deliverable may also be the implementation of (a part of) an architecture.

> A deliverable is defined as a precisely-defined outcome of a work package.

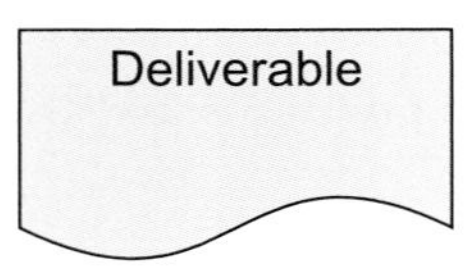

Figure 75: Deliverable Notation

Example

In PRINCE2, the deliverables (products) of a project are leading. The overall result of a project is described in a "project product description"; the hierarchical decomposition of this product in sub-products is shown in a Product Breakdown Structure, an example of which is shown in the model below.

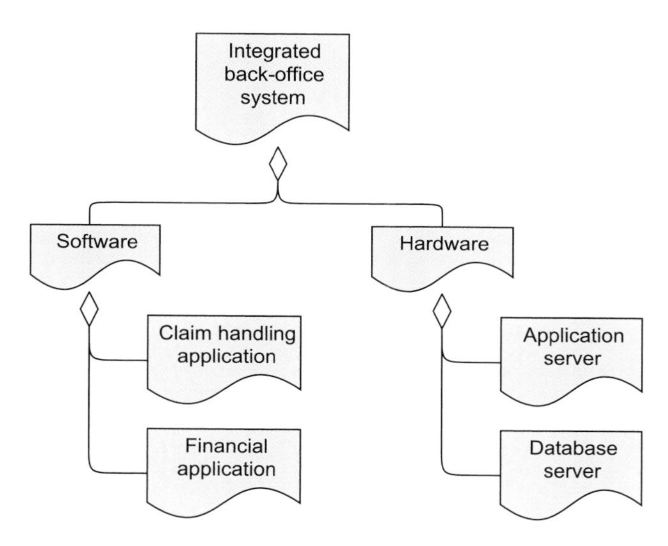

Example 58: Deliverable

11.2.3 Plateau

An important premise in TOGAF is that the various architectures are described for different stages in time. In each of the Phases B, C, and D of the ADM, a Baseline Architecture and Target Architecture are created, describing the current situation and the desired future situation. In Phase E (Opportunities and Solutions), so-called Transition Architectures are defined, showing the enterprise at incremental states reflecting periods of transition between the Baseline and Target Architectures. Transition Architectures are used to allow for individual work packages and projects to be grouped into managed portfolios and programs, illustrating the business value at each stage.

In order to support this, we introduce the *plateau* concept.

A plateau is defined as a relatively stable state of the architecture that exists during a limited period of time.

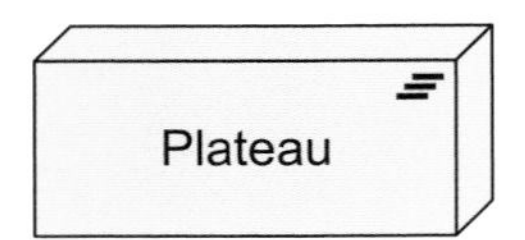

Figure 76: Plateau Notation

Example

The model below illustrates the use of the plateau concept to model the migration from Baseline to Target Architecture, defining a number of intermediate (possibly alternative) Transition Architectures.

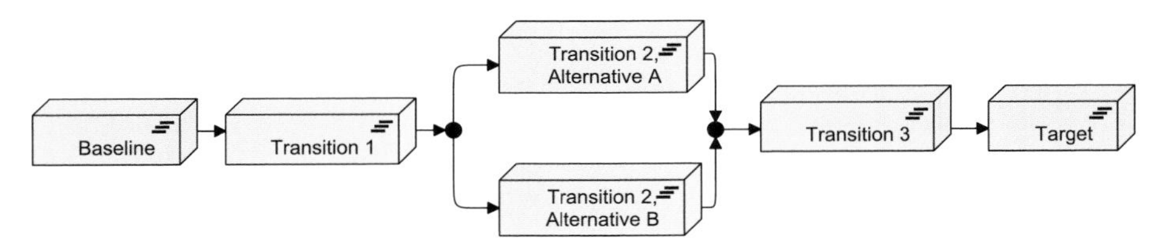

Example 59: Plateau

11.2.4 Gap

A *gap* is an important outcome of a gap analysis in Phases B, C, and D of the TOGAF ADM, and forms an important input for the subsequent implementation and migration planning. The gap concept is linked to two plateaus (e.g., Baseline and Target Architecture, or two subsequent Transition Architectures), and represents the differences between these plateaus.

A gap is defined as an outcome of a gap analysis between two plateaus.

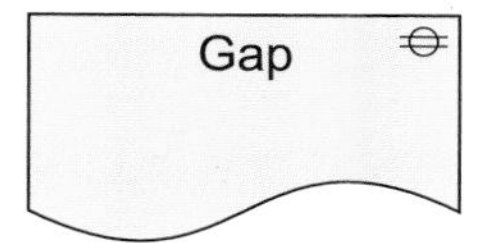

Figure 77: Gap Notation

> **Example**
> The model below illustrates the gap between the Baseline and Target infrastructure, showing which of the elements of the infrastructure are added to or removed from the Baseline.

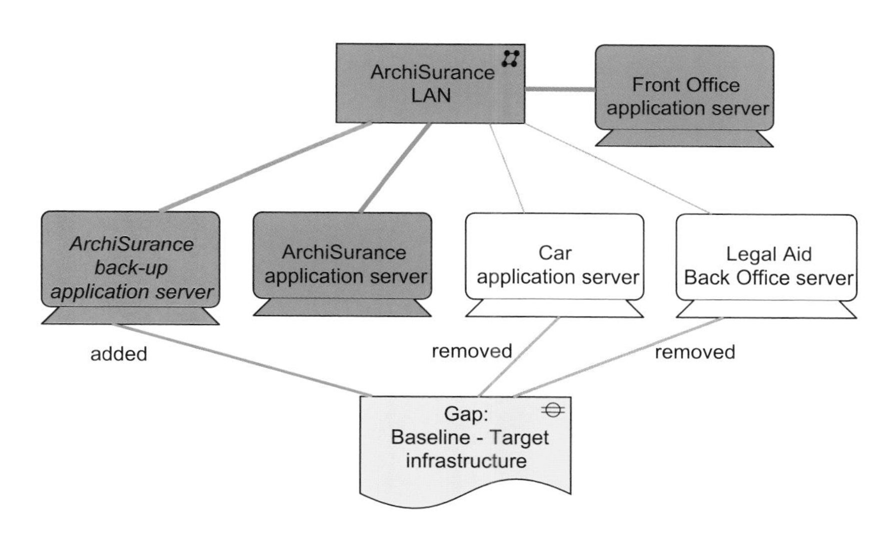

Example 60: Gap

11.2.5 Summary of Implementation and Migration Concepts

gives an overview of the implementation and migration concepts, with their definitions.

Table 34: Motivational Concepts

Concept	Definition	Notation
Work Package	A series of actions designed to accomplish a unique goal within a specified time.	Work package
Deliverable	A precisely-defined outcome of a work package.	Deliverable
Plateau	A relatively stable state of the architecture that exists during a limited period of time.	Plateau
Gap	An outcome of a gap analysis between two plateaus.	Gap

11.3 Relationships

The Implementation and Migration extension re-uses the standard ArchiMate relationships.

11.4 Cross-Aspect Dependencies

Figure 78 shows how the implementation and migration concepts can be related to the ArchiMate core concepts.

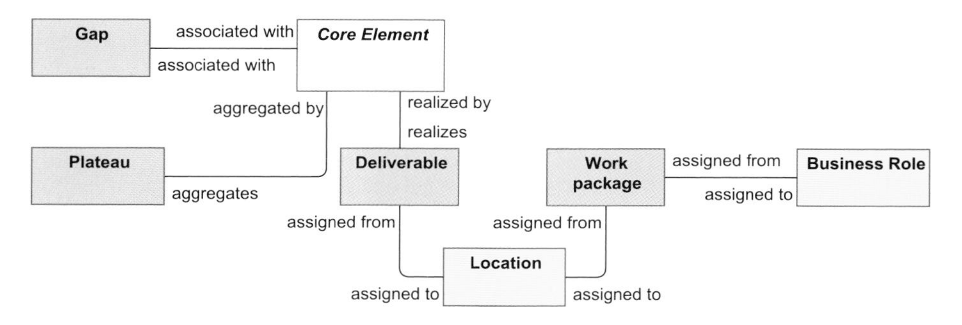

Figure 78: Relationships between Implementation & Migration Extension and the ArchiMate Core Concepts

A business role may be assigned to a work package.

A plateau is linked to an architecture that is valid for a certain time span. To indicate which parts of the architecture belong to a certain plateau, a plateau may aggregate any of the concepts of the ArchiMate core.

A gap is associated with the core concepts that are unique to one of the plateaus linked by the gap; i.e., the core concepts that make up the difference between these plateaus.

A deliverable may realize, among others, the implementation of an architecture or a part of an architecture. Therefore, any of the concepts of the ArchiMate core may be linked to a deliverable by means of a realization relationship.

Like most of the core concepts, a location may be assigned to a work package or deliverable.

Weaker relationships may also be defined. For example, the association relationship may be used to show that parts of the architecture are affected in some way by certain work packages.

Strictly speaking, the relationships between the implementation and migration concepts and the motivation concepts are indirect relationships; e.g., a deliverable realizes a requirement or goal through the realization of an ArchiMate core element (e.g., an application component, business process, or service). However, it is still useful to make these relationships explicit, to show directly that a deliverable is needed to realize certain requirements and goals.

Also, goals and requirements can be associated with a certain plateau; e.g., certain requirements may only be applicable to the Target Architecture, while others may apply to a certain Transition Architecture. This can be modeled by means of the aggregation relationship.

Figure 79 summarizes the relationships between the concepts of the Implementation and Migration extension and the concepts of the Motivation extension.

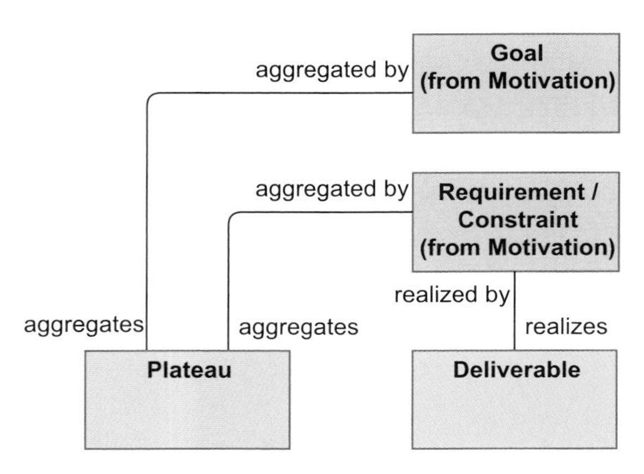

Figure 79: Relationships between Plateau, Project Result, and Motivation Concepts

11.5 Viewpoints

The following standard viewpoints for modeling implementation and migration aspects are distinguished:

- The *project viewpoint* is primarily used to model the management of architecture change.
- The *migration viewpoint* is used to model the transition from an existing architecture to a target architecture.
- The *implementation and migration viewpoint* is used to model the relationships between the programs and projects and the parts of the architecture that they implement.

All viewpoints are described separately below. For each viewpoint the comprised concepts and relationships, the guidelines for the viewpoint use, and the goal and target group and of the viewpoint are indicated. Furthermore, each viewpoint description contains example models. For more details on the goal and use of viewpoints, refer to [2], Chapter 8.

11.5.1 Project Viewpoint

A *project viewpoint* is primarily used to model the management of architecture change. The "architecture" of the migration process from an old situation (current state enterprise architecture) to a new desired situation (target state enterprise architecture) has significant consequences on the medium and long-term growth strategy and the subsequent decision-making process. Some of the issues that should be taken into account by the models designed in this viewpoint are:

- Developing fully-fledged organization-wide enterprise architecture is a task that may require several years.
- All systems and services must remain operational regardless all the presumable modifications and changes of the enterprise architecture during the change process.
- The change process may have to deal with immature technology standards (e.g., messaging, security, data, etc.).
- The change has serious consequences for the personnel, the culture, the way of working, and the organization.

Furthermore, there are several other governance aspects that might constrain the transformation process, such as internal and external co-operation, project portfolio management, project management (deliverables, goals, etc.), plateau planning, financial and legal aspects, etc.

Table 35: Description of the Project Viewpoint

Project Viewpoint	
Stakeholders	(operational) managers, enterprise and ICT architects, employees, shareholders
Concerns	Architecture vision and policies, motivation
Purpose	Deciding, informing
Abstraction Level	Overview
Layers/ Extensions	Implementation and Migration extension
Aspects	Information, behavior, structure

Concepts and Relationships

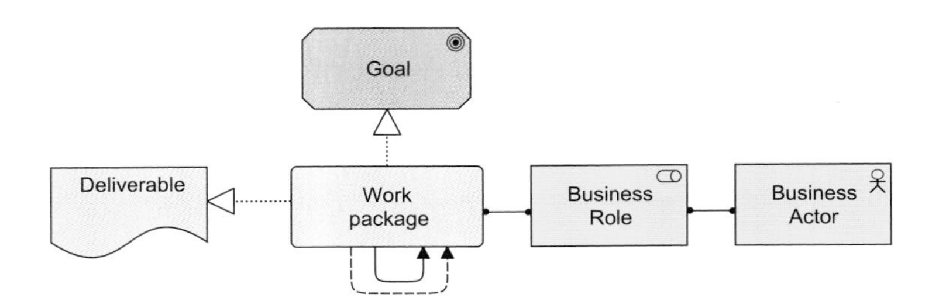

Example

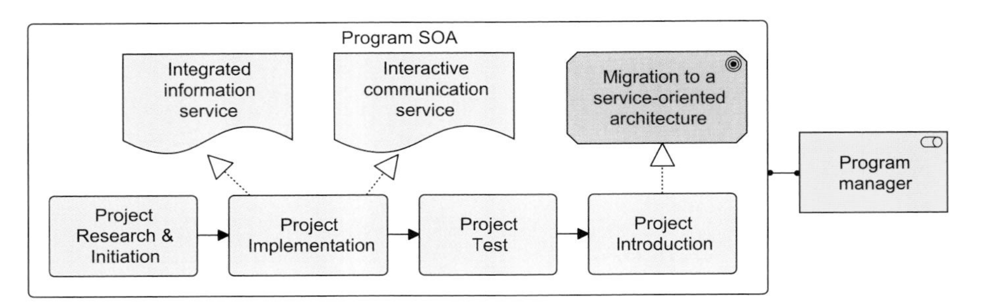

11.5.2 Migration Viewpoint

The *migration viewpoint* entails models and concepts that can be used for specifying the transition from an existing architecture to a desired architecture. Since the plateau and gap concepts have been quite extensively presented in Section 10.2, here the migration viewpoint is only briefly described and positioned by means of the table below.

Table 36: Description of the Migration Viewpoint

Migration Viewpoint	
Stakeholders	Enterprise architects, process architects, application architects, infrastructure architects and domain architects, employees, shareholders
Concerns	History of models
Purpose	Designing, deciding, informing
Abstraction Level	Overview
Layers/ Extensions	Implementation and Migration extension
Aspects	Not applicable.

Concepts and Relationships

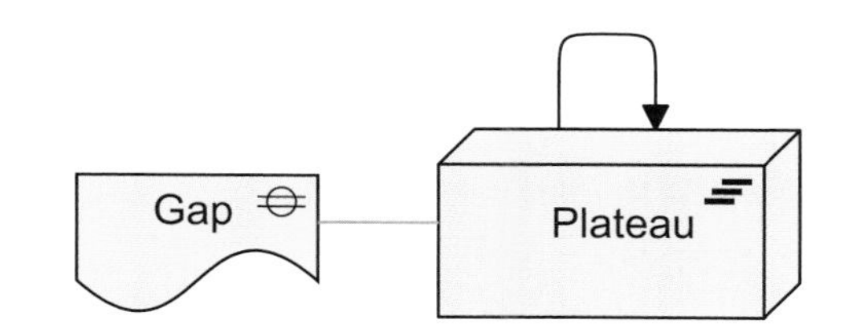

Example

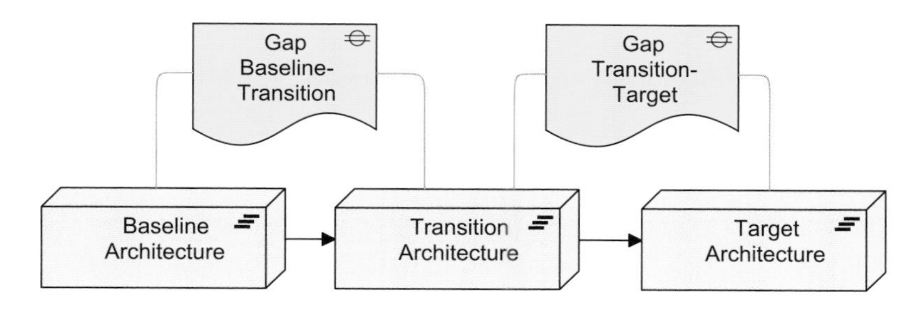

11.5.3 Implementation and Migration Viewpoint

The *implementation and migration viewpoint* is used to relate programs and projects to the parts of the architecture that they implement. This view allows modeling of the scope of programs, projects, project activities in terms of the plateaus that are realized or the individual architecture elements that are affected. In addition, the way the elements are affected may be indicated by annotating the relationships.

Furthermore, this viewpoint can be used in combination with the programs and projects viewpoint to support portfolio management:

- The programs and projects viewpoint is suited to relate business goals to programs and projects. For example, this makes it possible to analyze at a high level whether all business goals are covered sufficiently by the current portfolio(s).
- The implementation and migration viewpoint is suited to relate business goals (and requirements) via programs and projects to (parts of) the architecture. For example, this makes it possible to analyze potential overlap between project activities or to analyze the consistency between project dependencies and dependencies among plateaus or architecture elements.

Table 37: Description of the Architecture Implementation and Migration Viewpoint

Architecture Implementation and Migration Viewpoint	
Stakeholders	(operational) managers, enterprise and ICT architects, employees, shareholders
Concerns	Architecture vision and policies, motivation
Purpose	Deciding, informing
Abstraction Level	Overview
Layers/ Extensions	Business layer, application layer, technology layer, implementation & migration extension
Aspects	Information, behaviour, structure

Concepts and Relationships

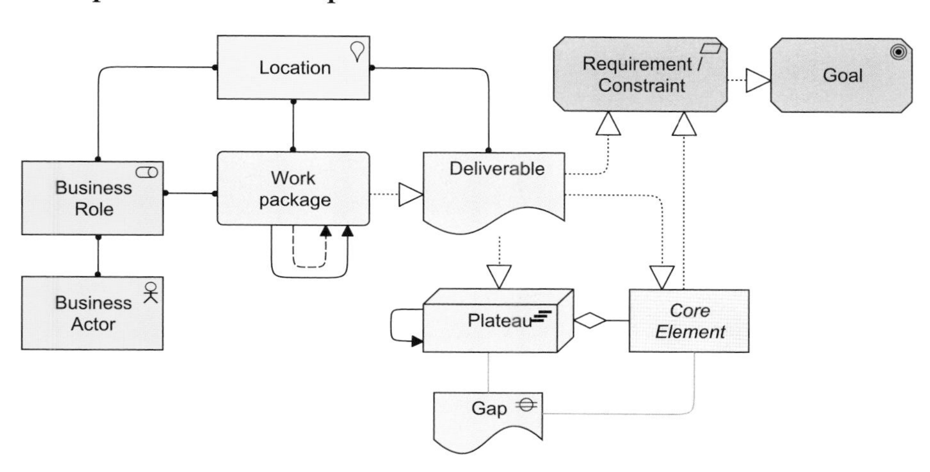

Example

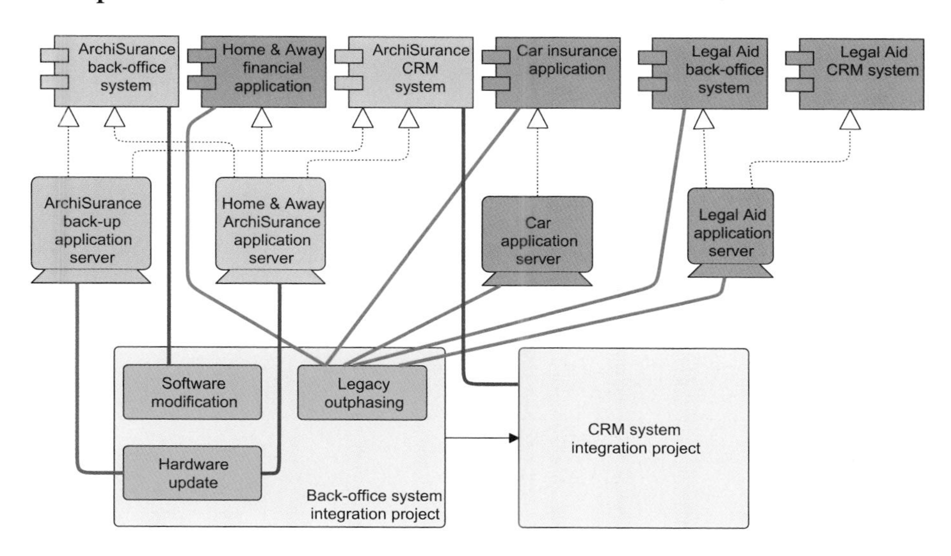

Chapter 12

Future Directions (Informative)

This chapter is informative. It should be used as a guide to current thinking; there is not necessarily a commitment to implement all of these future directions in their entirety.

The first version of the ArchiMate language as specified in Issue 1.0 of this Technical Standard has a strong focus on describing the operational aspects of an enterprise. In addition to this, the current version includes two extensions: the Motivation extension and the Implementation and Migration extension.

Although the aim is to keep the core of the language relatively small, a number of other directions for extending the language, as well as more advanced tool support for inherent features of ArchiMate models, can be envisaged. In this chapter, we identify some likely extensions for future versions of the language and associated tool support.

12.1 Extending and Refining the Concepts

In the practical use of ArchiMate, a number of areas have been identified in which a future extension of the language may be considered:

- Business policies and rules
- The design process
- Architecture-level predictions

Furthermore, there are a number of individual concepts that may be considered for future versions of the language; e.g.:

- Capability, defined as a collection of business and IT resources that together provide the ability to execute one or more business processes
- Milestone (as part of the Implementation & Migration extension)

12.1.1 Business Policies and Rules

Business policies are sets of general rules followed by a business that define business processes and practices. Business rules make these policies actionable for specific situations. Business rules separate business knowledge, based on, for example, legislation and regulations, business strategy, and business policies, from the business processes and systems that use this knowledge.

At the enterprise architecture level, sets of policies or rules may be modeled and linked to other elements of the architecture, such as business processes, application components, or services.

12.1.2 Design Process

Second, the language could provide additional support for the early stages of the architecture development process. In these early stages, architects often use informal, sketchy, and incomplete models that later evolve into formally correct ArchiMate models. Hence, a relaxation of formal correctness criteria in the early design stages might be in order. Support for this design evolution is closely related to the concepts from the Motivation extension, since design decisions are guided by goals, principles, and requirements, and the design process is instrumental to the evolution of the architecture.

12.1.3 Other Improvements

Next to the extensions in the areas mentioned above, some definitions of language concepts might also be improved and clarified. For example, the grouping concept could be given more explicit semantics. In practical use, some concepts have been used to good effect for other purposes than strictly intended; their future definitions may be updated to account for such usage.

A more formal specification of the metamodel of the language, expressed in a standard such as OMG's MOF or Encore (part of the Eclipse Modeling Framework), would facilitate the implementation of the language in software tools.

A

Summary of Language Notation

A.1 Core Concepts and Relationships

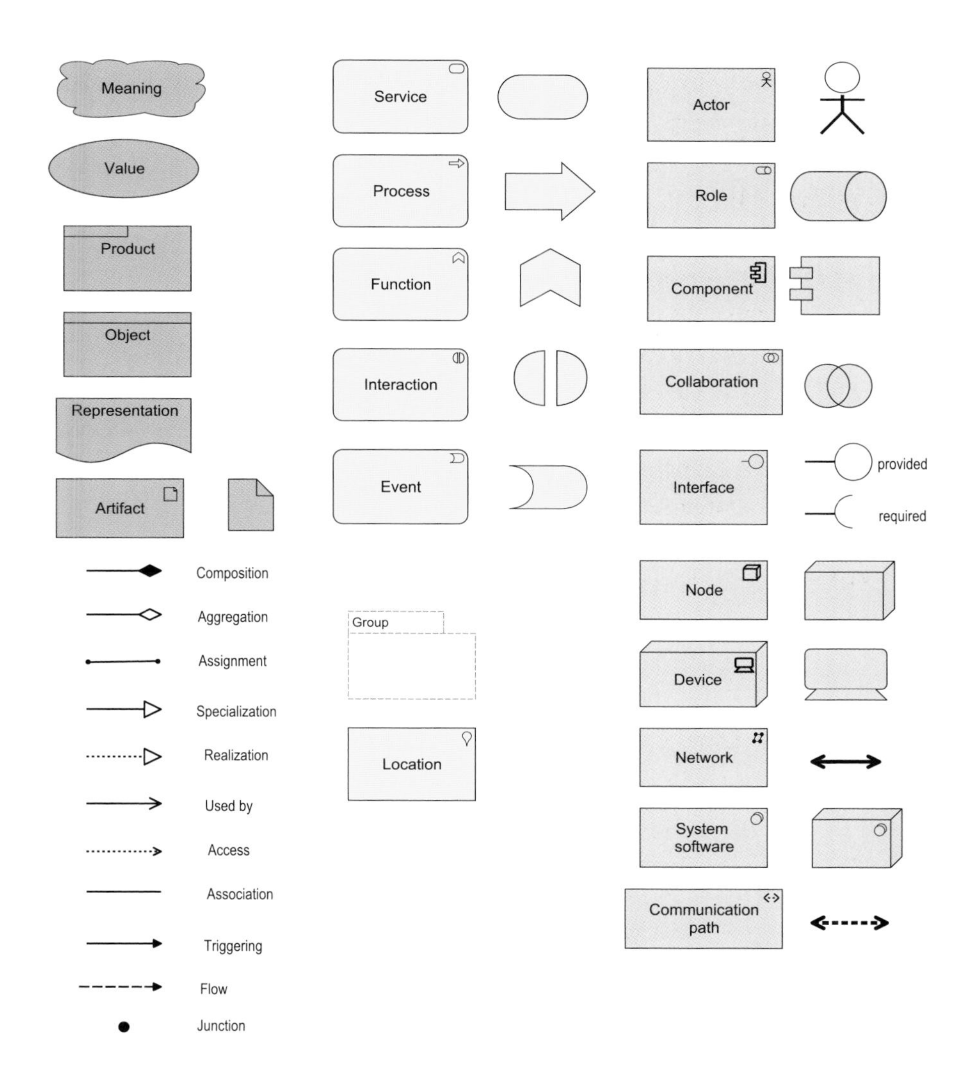

A.2 Extensions

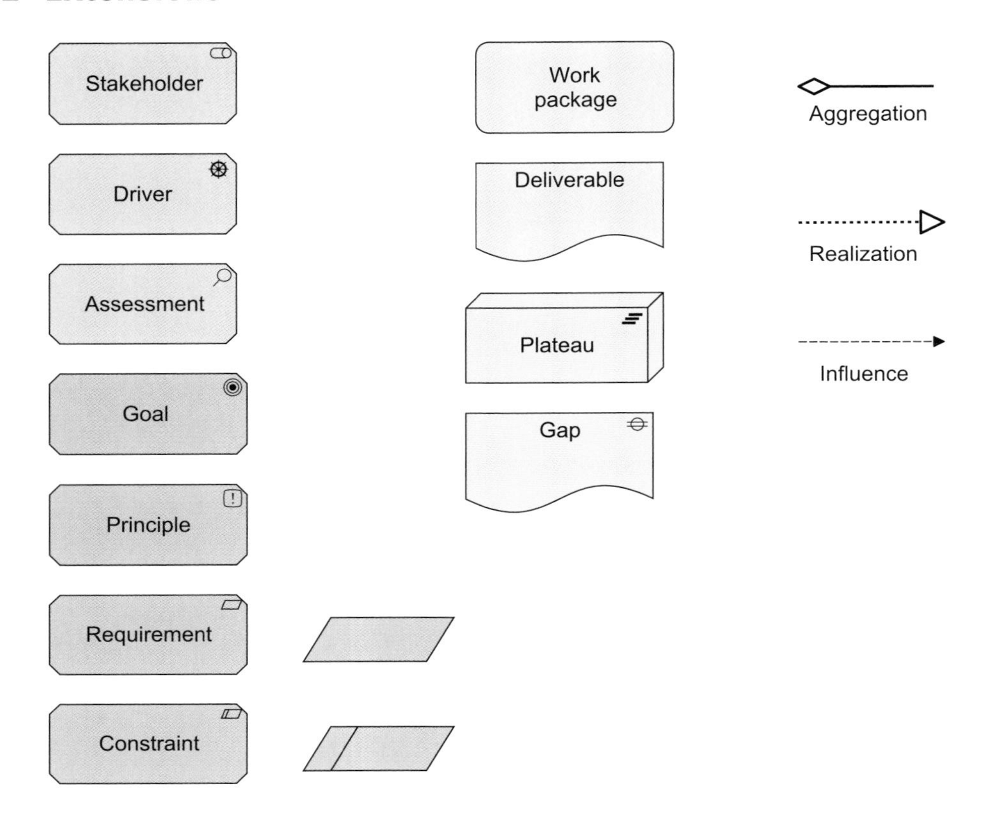

B

Overview of Relationships

B.1 Core Concepts

From ⇣ / To ⇢	Business Actor	Business Role	Business Collaboration	Location	Business Interface	Business Process	Business Function	Business Interaction	Business Event	Business Service	Business Object	Representation	Product	Contract	Meaning	Value
Business Actor	cfgostu	fiotu	fiotu	o	fiotu	fiotu	fiotu	fiotu	ot	ioru	ao	o	o	ao	o	o
Business Role	fotu	cfgostu	cfgostu	o	cfgiotu	fiotu	fiotu	fiotu	ot	ioru	ao	o	o	ao	o	o
Business Colla-boration	fgotu	cfgostu	cfgostu	o	cfgotu	iou	fiou	iou	ot	ioru	ao	o	o	ao	o	o
Location	io	io	io	cgos	io	io	io	io	io	io	io	io	o	io	o	o
Business Interface	fotu	fotu	fotu	o	cfgostu	ou	ou	ou	ot	iou	ao	o	o	ao	o	o
Business Process	fotu	fotu	fotu	o	fotu	cfgostu	cfgostu	cfgostu	ot	oru	ao	o	o	ao	o	o
Business Function	fotu	fotu	fotu	o	fotu	cfgostu	cfgostu	cfgostu	ot	oru	ao	o	o	ao	o	o
Business Interaction	fotu	fotu	fotu	o	fotu	cfgostu	cfgostu	cfgostu	ot	oru	ao	o	o	ao	o	o
Business Event	ot	ot	ot	ot	ot	ot	ot	ot	cgost	o	ao	o	o	ao	o	o
Business Service	ou	ou	ou	o	ou	ou	ou	ou	o	cfgostu	ao	o	o	ao	o	o
Business Object	o	o	o	o	o	o	o	o	o	o	cgos	o	o	cgos	o	o

From ↓ / To →	Business Actor	Business Role	Business Collaboration	Location	Business Interface	Business Process	Business Function	Business Interaction	Business Event	Business Service	Business Object	Representation	Product	Contract	Meaning	Value
Representation	o	o	o	o	o	o	o	o	o	o	or	cgos	o	or	o	o
Product	ou	ou	ou	o	ou	ou	ou	ou	o	gou	ao	o	cgos	ago	o	o
Contract	o	o	o	o	o	o	o	o	o	o	cgos	o	o	cgos	o	o
Meaning	o	o	o	o	o	o	o	o	o	o	o	o	o	o	cgos	o
Value	o	o	o	o	o	o	o	o	o	o	o	o	o	o	o	cgos
Application Component	fotu	fotu	fotu	o	fotu	fiotu	iou	iou	o	ioru	ao	o	o	ao	o	o
Application Collaboration	fotu	fotu	fotu	o	fotu	fiotu	fiotu	iou	o	ioru	ao	o	o	ao	o	o
Application Interface	ou	ou	ou	o	fotu	fotu	fotu	ou	o	ou	ao	o	o	ao	o	o
Application Function	ou	ou	ou	o	fotu	fotu	ou	ou	o	ou	ao	o	o	ao	o	o
Application Interaction	ou	ou	ou	o	fotu	fotu	ou	ou	o	ou	ao	o	o	ao	o	o
Application Service	ou	ou	ou	o	ou	ou	ou	ou	o	ou	ao	o	o	ao	o	o
Data Object	o	o	o	o	o	o	o	o	o	o	or	o	o	or	o	o
Node	ou	ou	ou	o	ou	oru	oru	oru	o	oru	aoru	o	o	aoru	o	o
Device	ou	ou	ou	o	ou	oru	oru	oru	o	oru	aoru	o	o	aoru	o	o

From / To	Business Actor	Business Role	Business Collaboration	Location	Business Interface	Business Process	Business Function	Business Interaction	Business Event	Business Service	Business Object	Representation	Product	Contract	Meaning	Value
System Software	ou	ou	ou	o	ou	oru	oru	oru	o	oru	aoru	o	o	aoru	o	o
Infrastructure Interface	aou	aou	aou	o	aou	aou	aou	aou	o	aou	aou	o	o	aou	o	o
Network	o	o	o	o	o	o	o	o	o	o	o	o	o	o	o	o
Communication Path	o	o	o	o	o	o	o	o	O	o	o	o	o	o	o	o
Infrastructure Function	ou	ou	ou	o	ou	ou	oru	oru	O	oru	aou	o	o	aou	o	o
Infrastructure Service	ou	ou	ou	o	ou	ou	ou	ou	O	ou	aou	o	o	aou	o	o
Artifact	ou	ou	ou	o	ou	oru	oru	oru	O	oru	aor	o	o	aor	o	o
Junction	ft	ft	ft	o	ft	ft	ft	ft	ft	ft						

Relationships

(a)ccess, ass(i)gnment, (c)omposition, (r)ealization, (t)riggering
a(g)gregation, ass(o)ciation, (f)low, (s)pecialization, (u)sed by

From ⇢ / To ⇢	Business Actor	Business Role	Business Collaboration	Location	Business Process	Business Function	Business Interaction	Business Event	Business Service	Business Object	Representation	Product	Contract	Meaning
Junction	ft	ft	ft	io	ft	ft	ft	ft	ft					
Artifact	o	o	o	io	o	o	o	o	o	o	o	ao	o	o
Infrastructure Service	o	o	o	io	o	o	o	o	o	o	o	gou	o	o
Infrastructure Function	o	o	o	io	o	o	o	o	o	o	o	ou	o	o
Communication Path	o	o	o	io	o	o	o	o	o	o	o	o	o	o
Network	o	o	o	io	o	o	o	o	o	o	o	o	o	o
Infrastructure Interface	o	o	o	io	o	o	o	o	o	o	o	ou	o	o
System Software	o	o	o	io	o	o	o	o	o	o	o	ou	o	o
Device	o	o	o	io	o	o	o	o	o	o	o	ou	o	o
Node	o	o	o	io	o	o	o	o	o	o	o	ou	o	o
Data Object	o	o	o	io	o	o	o	o	o	o	o	ao	o	o
Application Service	o	o	o	io	o	o	o	o	o	o	o	gou	o	o
Application Interaction	fot	fot	fot	io	o	o	o	o	o	o	o	o	o	o
Application Function	fot	fot	fot	io	o	o	o	o	o	o	o	ou	o	o
Application Interface	fot	fot	o	io	o	o	o	o	o	o	o	ou	o	o
Application Collaboration	fot	fot	fot	io	fot	fot	fot	ot	o	o	o	ou	o	o
Application Component	fot	fot	fot	io	fot	fot	fot	ot	o	o	o	ou	o	o

From ⇣ / To ⇢	Application Component	Application Collaboration	Application Interface	Application Function	Application Interaction	Application Service	Data Object	Node	Device	System Software	Infrastructure Interface	Network	Communication Path	Infrastructure Function	Infrastructure Service	Artifact	Junction
Value	o	o	o	o	o	o	o	o	o	o	o	o	o	o	o	o	
Application Component	cfgostu	cfgostu	cfgotu	iou	iou	ioru	ao	o	o	o	o	o	o	o	o	o	ft
Application Collaboration	cfgostu	cfgostu	cfgotu	iou	iou	ioru	ao	o	o	o	o	o	o	o	o	o	ft
Application Interface	fotu	fotu	cfgostu	ou	ou	iou	ao	o	o	o	o	o	o	o	o	o	ft
Application Function	ou	ou	ou	cfgostu	fot	oru	ao	o	o	o	o	o	o	o	o	o	ft
Application Interaction	ou	ou	ou	fotu	cfgost	oru	ao	o	o	o	o	o	o	o	o	o	ft
Application service	ou	ou	ou	ou	ou	cfgostu	ao	o	o	o	o	o	o	o	o	o	ft
Data Object	o	o	o	o	o	o	cgos	o	o	o	o	o	o	o	o	o	
Node	aoru	aoru	aoru	aoru	aoru	aoru	aoru	cfgostu	cfgostu	cfgostu	cfgotu	o	o	iou	ioru	aiou	ft
Device	aoru	aoru	aoru	aoru	aoru	aoru	aoru	cfgostu	cfgostu	cfgiostu	cfgotu	o	o	iou	ioru	aiou	ft
System Software	aoru	aoru	aoru	aoru	aoru	aoru	aoru	cfgostu	cfgostu	cfgostu	cfgotu	o	o	iou	ioru	aiou	ft
Infrastructure Interface	aou	aou	aou	aou	aou	aou	aou	fotu	fotu	fotu	cfgostu	o	o	ou	iou	aou	ft
Network	o	o	o	o	o	o	o	o	o	o	o	cgos	or		o	o	

From / To	Application Component	Application Collaboration	Application Interface	Application Function	Application Interaction	Application Service	Data Object	Node	Device	System Software	Infrastructure Interface	Network	Communication Path	Infrastructure Function	Infrastructure Service	Artifact	Junction
Communication Path	o	o	o	o	o	o	o	o	o	o	o	o	cgos	o	o	o	
Infrastructure Function	aou	aou	aou	aou	aou	auo	aou	ou	ou	ou	ou	o	c	cfgostu	oru	aou	ft
Infrastructure Service	ou	ou	ou	ou	ou	ou	aou	ou	ou	ou	ou	o	o	ou	cfgostu	aou	ft
Artifact	oru	oru	oru	oru	oru	oru	aor	o	o	or	or	o	o	or	or	cgors	
Junction	ft	ft	ft	ft	ft	ft		ft	ft	ft	ft			ft	ft		ft

Relationships

(a)ccess ass(i)gnment (c)omposition (r)ealization (t)riggering
a(g)gregation ass(o)ciation (f)low (s)pecialization (u)sed by

B.2 Extensions

From ⇣ / To ⇢	Stakeholder	Driver	Assessment	Goal	Requirement	Principle	Constraint	Work Package	Deliverable	Plateau	Gap	Core Element	Business Actor	Business Role	Location	Value
Stakeholder	gcso	o	o	o	o	o	o	o	o	o	o	o	o	o	o	
Driver	o	gcson	on	on	on	on	on	o	o	o	o	o	o	o	o	on
Assessment	o	on	gcson	on	on	on	on	o	o	o	o	o	o	o	o	on
Goal	o	on	on	gcson	on	on	on	o	o	o	o	o	o	o	o	on
Requirement	o	on	on	ron	gcson	ron	gcson	o	o	o	o	o	o	o	o	on
Principle	o	on	on	ron	on	gcson	on	o	o	o	o	o	o	o	o	on
Constraint	o	on	on	ron	gcson	ron	gcson	o	o	o	o	o	o	o	o	o
Work Package	o	o	o	ro	ro	o	ro	gcsoft	ro	ro	o	ro	roft	roft	o	o
Deliverable	o	o	o	ro	ro	ro	ro	o	gcso	o	o	go	go	go	o	o
Plateau	o	o	o	gro	gro	o	gro	o	o	gcsot	o	go	go	go	o	o
Gap	o	o	o	o	o	o	o	o	o	o	gcso	o	o	o	o	o
Core Element	o	o	o	ro	ro	ro	ro	o	o	o	o					
Business Actor	io	o	o	ro	ro	ro	ro	ioft	ro	o	o					
Business Role	o	o	o	ro	ro	ro	ro	ioft	ro	o	o					
Location	io	o	o	o	o	o	o	io	io	o	o					
Value	o	o	o	o	o	o	o	o	o	o	o					

Relationships

(c)omposition i(n)fluence ass(i)gnment ass(o)ciation (t)riggering
(f)low a(g)gregation (r)ealization (s)pecialization

Index